Renewable Energy for All

Making Clean Power Accessible and Affordable

Lulu Pope

Renewable Energy for All

TABLE OF CONTENTS

Chapter 1: Solar Power for Everyone: Democratizing Sunlight

Basics of Solar Energy and Its Potential for Accessibility

Solar energy, a cornerstone of renewable energy sources, harnesses the sun's power to generate electricity and heat. This abundant and sustainable resource offers immense potential for accessibility, particularly in regions where traditional energy infrastructure is lacking or unreliable. Understanding the basics of solar energy is crucial for recognizing its potential to democratize energy access and empower communities worldwide.

At its core, solar energy relies on photovoltaic (PV) cells, which convert sunlight directly into electricity. These cells are made from semiconductor materials, such as silicon, that absorb photons from sunlight, releasing electrons and generating an electric current. The efficiency of PV cells has improved significantly over the years, making solar energy a viable option for a wide range of applications, from small-scale residential systems to large solar farms.

One of the most compelling aspects of solar energy is its scalability. Solar panels can be installed on rooftops, integrated into building materials, or deployed in vast arrays in open fields. This flexibility allows for tailored solutions that meet the specific needs of different communities. For instance, in urban areas, rooftop solar installations can provide a decentralized energy source, reducing reliance on centralized power grids and lowering electricity costs for residents. In rural or remote areas, standalone solar systems can offer a reliable source of power where traditional grid connections are impractical or too costly.

The potential for solar energy to enhance accessibility is further amplified by the decreasing cost of solar technology. Over the past decade, the price of solar panels has plummeted, driven by advancements in manufacturing processes and economies of scale. This trend has made solar energy more affordable for individuals and communities, paving the way for widespread adoption. Additionally, innovative financing models, such as solar leasing and power purchase agreements, have emerged to lower the upfront costs of solar installations, making them accessible to a broader audience.

Community solar projects represent another avenue for expanding solar energy access. These initiatives allow multiple participants to invest in or subscribe to a shared solar array, receiving credits on their electricity bills for the power generated. Community solar projects are particularly beneficial for individuals who cannot install solar panels on their properties, such as renters or those with shaded roofs. By pooling resources and sharing the benefits, community solar projects democratize access to solar energy and foster a sense of collective ownership and responsibility.

Despite the promising potential of solar energy, several challenges must be addressed to ensure its accessibility for all. Financial barriers remain a significant hurdle, particularly for low-income households and communities. While the cost of solar technology has decreased, the initial investment can still be prohibitive for some. To overcome this obstacle, governments and organizations can implement policies and programs that provide financial incentives, such as tax credits, rebates, or grants, to support solar adoption. Additionally, targeted outreach and education efforts can raise awareness about the benefits of solar energy and available resources, empowering individuals to make informed decisions.

Technological innovations also play a crucial role in enhancing the accessibility of solar energy. Researchers and engineers are continually developing new materials and designs to improve the efficiency and affordability of solar panels. For example, thin-film solar cells, which use less material and are lighter and more flexible than traditional silicon-based panels, offer promising potential for reducing costs and expanding applications. Similarly, advancements in energy storage technologies, such as batteries, can enhance the reliability of solar systems by storing excess energy for use during periods of low sunlight.

Case studies of successful solar initiatives in low-income areas provide valuable insights into the potential of solar energy to transform communities. In some regions, solar-powered microgrids have been established to provide electricity to off-grid villages, enabling access to essential services such as lighting, refrigeration, and communication. These projects not only improve the quality of life for residents but also stimulate economic development by supporting local businesses and creating job opportunities.

In urban settings, solar energy has been integrated into affordable housing projects, reducing energy costs for residents and promoting environmental sustainability. These initiatives often involve partnerships between government agencies, non-profit organizations, and private companies, highlighting the importance of collaboration in advancing solar accessibility.

The environmental benefits of solar energy further underscore its potential as a sustainable and accessible energy source. Unlike fossil fuels, solar power generates electricity without emitting greenhouse gases or other pollutants, contributing to cleaner air and a healthier environment. By reducing reliance on non-renewable energy sources, solar energy can play a pivotal role in mitigating climate change and promoting energy independence.

In conclusion, the basics of solar energy reveal a powerful and versatile resource with the potential to enhance energy accessibility for communities worldwide. Through technological advancements, innovative financing models, and collaborative efforts, solar energy can be harnessed to provide clean, affordable, and reliable power to those who need it most. As we continue to explore and expand the possibilities of solar energy, its role in democratizing energy access and fostering sustainable development becomes increasingly clear.

Community Solar Projects: Sharing the Sun's Power

Community solar projects represent a transformative approach to harnessing solar energy, offering a shared solution that democratizes access to renewable power. These projects allow multiple individuals or entities to invest in or subscribe to a single solar array, typically located off-site. Participants receive credits on their electricity bills for the power generated, effectively sharing the benefits of solar energy without the need for individual installations. This model is particularly advantageous for those who cannot install solar panels on their properties, such as renters, apartment dwellers, or homeowners with shaded roofs.

The concept of community solar is rooted in the idea of collective ownership and responsibility. By pooling resources, participants can achieve economies of scale, reducing the overall cost of solar energy and making it more accessible to a broader audience. This collaborative approach not only lowers financial barriers but also fosters a sense of community and shared purpose, as individuals come together to support a sustainable energy future.

One of the key benefits of community solar projects is their ability to provide clean, renewable energy to underserved populations. In many cases, low-income households face significant obstacles to accessing solar power, including high upfront costs and limited financing options. Community solar projects address these challenges by offering flexible subscription models and financing arrangements, allowing participants to pay for their share of the solar array over time. This approach makes solar energy more affordable and accessible, empowering individuals to reduce their carbon footprint and lower their energy bills.

The success of community solar projects often hinges on effective partnerships between various stakeholders, including government agencies, non-profit organizations, utilities, and private companies. These collaborations are essential for navigating the complex regulatory landscape and securing the necessary funding and resources to bring projects to fruition. Government incentives and policies, such as tax credits, grants, and renewable energy mandates, play a crucial role in supporting community solar initiatives and encouraging widespread adoption.

In addition to financial and regulatory considerations, the design and implementation of community solar projects must also take into account the unique needs and preferences of the participating community. Engaging with local residents and stakeholders early in the planning process is vital for building trust and ensuring that the project aligns with the community's values and priorities. This participatory approach can help identify potential challenges and opportunities, such as site selection, project size, and subscription models, ultimately leading to more successful and sustainable outcomes.

Community solar projects also offer significant environmental benefits, contributing to the reduction of greenhouse gas emissions and the transition to a cleaner energy grid. By displacing fossil fuel-based electricity with renewable solar power, these projects help mitigate climate change and improve air quality, benefiting both local communities and the broader environment. Moreover, community solar projects can serve as a catalyst for further renewable energy development, inspiring other communities to pursue similar initiatives and fostering a culture of sustainability.

The potential for community solar projects to drive economic development should not be overlooked. These projects create jobs in the solar industry, from manufacturing and installation to maintenance and administration. By investing in local labor and resources, community solar projects can stimulate economic growth and provide valuable workforce training and development opportunities. Additionally, the cost savings generated by community solar can free up household income for other essential needs, further enhancing the economic well-being of participants.

Despite the numerous advantages of community solar projects, several challenges must be addressed to ensure their success and scalability. One of the primary obstacles is the complexity of navigating the regulatory and policy landscape, which can vary significantly between jurisdictions. Streamlining permitting processes and standardizing regulations can help reduce barriers to entry and facilitate the development of community solar projects. Additionally, ongoing education and outreach efforts are essential for raising awareness about the benefits of community solar and encouraging participation among diverse populations.

Innovative business models and technological advancements also hold promise for expanding the reach and impact of community solar projects. For example, virtual net metering allows participants to receive credits for their share of the solar array's output, even if they are not physically connected to the grid. This approach can enhance the flexibility and accessibility of community solar, enabling more individuals to participate regardless of their location or utility provider. Similarly, advancements in energy storage technologies can improve the reliability and resilience of community solar projects, ensuring a consistent supply of power even during periods of low sunlight.

Case studies of successful community solar projects offer valuable insights and lessons for future initiatives. In some regions, community solar projects have been integrated into affordable housing developments, providing residents with access to clean energy and reducing their overall energy costs. These projects often involve partnerships between housing authorities, non-profit organizations, and solar developers, highlighting the importance of collaboration in achieving shared goals. Other successful examples include community solar projects that prioritize participation from low-income households, ensuring that the benefits of solar energy are equitably distributed.

The future of community solar projects is bright, with significant potential for growth and innovation. As more communities recognize the value of shared solar solutions, the demand for community solar projects is expected to increase, driving further investment and development in the sector. By continuing to explore new models and technologies, and by fostering collaboration among diverse stakeholders, community solar projects can play a pivotal role in advancing renewable energy access and promoting a more sustainable and equitable energy future.

Innovations in Low-Cost Solar Technologies

Solar technology has undergone remarkable advancements, making it more accessible and affordable for a diverse range of users. Innovations in low-cost solar technologies have played a crucial role in this transformation, enabling wider adoption and integration into everyday life. These innovations not only reduce the financial barriers associated with solar energy but also enhance its efficiency and versatility, paving the way for a sustainable energy future.

One of the most significant breakthroughs in low-cost solar technology is the development of thin-film solar cells. Unlike traditional silicon-based photovoltaic cells, thin-film cells are made by depositing one or more layers of photovoltaic material onto a substrate, such as glass, plastic, or metal. This process requires less material and energy, resulting in lower production costs. Thin-film solar cells are also lightweight and flexible, making them ideal for a variety of applications, from portable solar chargers to building-integrated photovoltaics. Their adaptability allows for creative installations on surfaces that were previously unsuitable for solar panels, expanding the potential for solar energy generation.

Another promising innovation is the use of perovskite materials in solar cells. Perovskites are a class of materials with a unique crystal structure that has shown exceptional potential for photovoltaic applications. Perovskite solar cells have achieved impressive efficiency gains in a relatively short period, rivaling traditional silicon cells. Moreover, they can be manufactured using low-cost, solution-based processes, further reducing production expenses. Researchers are actively exploring ways to improve the stability and scalability of perovskite solar cells, with the goal of bringing them to market as a cost-effective alternative to conventional solar technologies.

The integration of solar energy with other technologies has also contributed to cost reductions and increased accessibility. For instance, solar panels combined with energy storage systems, such as batteries, allow users to store excess energy generated during the day for use at night or during periods of low sunlight. This capability enhances the reliability and resilience of solar power, making it a more attractive option for both residential and commercial users. Advances in battery technology, including the development of more efficient and affordable lithium-ion and solid-state batteries, have further bolstered the appeal of solar-plus-storage solutions.

Microinverters and power optimizers represent another area of innovation that has improved the performance and cost-effectiveness of solar systems. Traditional solar installations use a single, centralized inverter to convert the direct current (DC) generated by solar panels into alternating current (AC) for use in homes and businesses. However, this approach can lead to inefficiencies, particularly if one or more panels are shaded or malfunctioning. Microinverters and power optimizers address this issue by optimizing the output of individual panels, maximizing energy production and reducing losses. These technologies have become increasingly affordable, making them a viable option for enhancing the efficiency of solar installations.

The rise of do-it-yourself (DIY) solar kits has also contributed to the democratization of solar energy. These kits provide all the necessary components for individuals to install their own solar systems, often at a fraction of the cost of professional installations. DIY solar kits are particularly popular in rural or off-grid areas, where access to professional installation services may be limited. By empowering individuals to take control of their energy needs, DIY solar kits promote energy independence and self-sufficiency.

Innovations in solar financing have played a pivotal role in making solar technology more accessible. Traditional financing models often require significant upfront investments, which can be a barrier for many potential users. To address this challenge, new financing options, such as solar leasing and power purchase agreements (PPAs), have emerged. These models allow users to install solar systems with little or no upfront cost, paying for the energy generated over time. This approach reduces financial risk and makes solar energy more attainable for a broader audience.

Community solar projects, which allow multiple participants to share the benefits of a single solar installation, have also gained traction as a cost-effective way to expand solar access. By pooling resources, participants can achieve economies of scale, reducing the overall cost of solar energy and making it more affordable for everyone involved. These projects are particularly beneficial for individuals who cannot install solar panels on their properties, such as renters or those with shaded roofs.

The role of policy and regulation in supporting low-cost solar innovations cannot be overstated. Government incentives, such as tax credits, rebates, and renewable energy mandates, have been instrumental in driving the adoption of solar technology. By creating a favorable regulatory environment, policymakers can encourage investment in research and development, leading to further advancements in solar technology and cost reductions.

Education and outreach efforts are also essential for promoting the adoption of low-cost solar technologies. By raising awareness about the benefits of solar energy and the availability of affordable options, individuals and communities can make informed decisions about their energy needs. Workshops, training programs, and informational resources can empower users to take advantage of the latest innovations in solar technology, fostering a culture of sustainability and environmental stewardship.

The future of low-cost solar technology is bright, with ongoing research and development promising even greater advancements. As new materials, manufacturing processes, and business models continue to emerge, the cost of solar energy is expected to decline further, making it an increasingly attractive option for a wide range of users. By embracing these innovations, we can accelerate the transition to a clean energy future, ensuring that the benefits of solar power are accessible to all.

Overcoming Financial Barriers to Solar Adoption

Financial barriers have long been a significant obstacle to the widespread adoption of solar energy. Despite the decreasing costs of solar technology, the initial investment required for installation can still be prohibitive for many individuals and communities. Overcoming these financial hurdles is crucial for making solar energy accessible to a broader audience and ensuring that its benefits are equitably distributed. By exploring various strategies and solutions, we can pave the way for more inclusive solar adoption.

One of the most effective ways to address financial barriers is through innovative financing models that reduce or eliminate upfront costs. Solar leasing and power purchase agreements (PPAs) have emerged as popular options for individuals and businesses looking to go solar without the burden of a large initial investment. Under these arrangements, a third-party company installs and maintains the solar system, while the user pays for the energy generated over time. This approach allows users to benefit from solar energy immediately, with predictable monthly payments that are often lower than their previous electricity bills.

Government incentives and subsidies also play a critical role in making solar energy more affordable. Tax credits, rebates, and grants can significantly reduce the cost of solar installations, making them more accessible to a wider range of users. For example, the federal Investment Tax Credit (ITC) in the United States allows homeowners and businesses to deduct a percentage of their solar installation costs from their taxes. Many states and local governments offer additional incentives, further lowering the financial barriers to solar adoption. By taking advantage of these programs, individuals and communities can make solar energy a more viable option.

Community solar projects offer another avenue for overcoming financial barriers. By pooling resources, participants can share the costs and benefits of a single solar installation, achieving economies of scale that reduce the overall expense. This model is particularly beneficial for individuals who cannot install solar panels on their properties, such as renters or those with shaded roofs. Community solar projects often involve flexible subscription models, allowing participants to pay for their share of the solar array over time. This approach makes solar energy more affordable and accessible, empowering individuals to reduce their carbon footprint and lower their energy bills.

Low-interest loans and financing programs specifically designed for solar installations can also help mitigate financial barriers. Many banks and credit unions offer green loans with favorable terms for renewable energy projects, enabling individuals to finance their solar systems with manageable monthly payments. Additionally, some utilities and non-profit organizations provide on-bill financing, allowing customers to pay for their solar installations through their utility bills. This approach simplifies the payment process and can make solar energy more attainable for those with limited access to traditional financing options.

Education and outreach efforts are essential for raising awareness about the financial resources available for solar adoption. By providing information about incentives, financing options, and the long-term benefits of solar energy, individuals and communities can make informed decisions about their energy needs. Workshops, training programs, and informational resources can empower users to take advantage of the latest innovations in solar technology, fostering a culture of sustainability and environmental stewardship.

Partnerships between government agencies, non-profit organizations, and private companies can also play a vital role in overcoming financial barriers to solar adoption. Collaborative efforts can lead to the development of targeted programs and initiatives that address the unique needs of different communities. For example, some programs focus on providing solar access to low-income households, ensuring that the benefits of solar energy are equitably distributed. By working together, stakeholders can leverage their resources and expertise to create more inclusive and sustainable energy solutions.

Technological advancements continue to drive down the cost of solar energy, making it more accessible to a broader audience. Innovations in solar panel manufacturing, such as the development of thin-film and perovskite solar cells, have reduced production costs and increased efficiency. These advancements, combined with improvements in energy storage and grid integration, enhance the overall value proposition of solar energy, making it a more attractive option for individuals and businesses alike.

The environmental and economic benefits of solar energy further underscore the importance of overcoming financial barriers. By reducing reliance on fossil fuels, solar power contributes to cleaner air and a healthier environment, benefiting both local communities and the broader ecosystem. Additionally, solar energy can stimulate economic growth by creating jobs in the solar industry, from manufacturing and installation to maintenance and administration. By investing in local labor and resources, solar projects can provide valuable workforce training and development opportunities, further enhancing the economic well-being of participants.

Case studies of successful solar initiatives offer valuable insights and lessons for overcoming financial barriers. In some regions, solar-powered microgrids have been established to provide electricity to off-grid villages, enabling access to essential services such as lighting, refrigeration, and communication. These projects not only improve the quality of life for residents but also stimulate economic development by supporting local businesses and creating job opportunities. In urban settings, solar energy has been integrated into affordable housing projects, reducing energy costs for residents and promoting environmental sustainability.

The future of solar energy is bright, with significant potential for growth and innovation. As new financing models, technologies, and policies continue to emerge, the cost of solar energy is expected to decline further, making it an increasingly attractive option for a wide range of users. By embracing these innovations and working collaboratively to address financial barriers, we can accelerate the transition to a clean energy future, ensuring that the benefits of solar power are accessible to all.

Case Studies: Successful Solar Initiatives in Low-Income Areas

In the heart of a bustling city, where skyscrapers cast long shadows over narrow streets, a remarkable transformation is taking place. A community once plagued by high energy costs and unreliable power supply is now basking in the glow of solar energy. This is the story of a successful solar initiative that has brought light and hope to a low-income neighborhood, demonstrating the profound impact of renewable energy on communities in need.

The project began with a simple yet ambitious goal: to provide affordable and reliable electricity to residents who had long struggled with energy insecurity. Spearheaded by a coalition of local non-profits, government agencies, and private companies, the initiative sought to harness the power of the sun to address the community's energy challenges. Through a combination of innovative financing, community engagement, and cutting-edge technology, the project has become a beacon of sustainability and empowerment.

At the heart of the initiative is a community solar array, strategically located on a previously underutilized plot of land. The solar panels, glistening under the sun, generate clean electricity that is distributed to participating households through a virtual net metering system. This setup allows residents to receive credits on their electricity bills for the power produced by the solar array, effectively reducing their energy costs without the need for individual rooftop installations.

One of the key factors contributing to the project's success is its inclusive financing model. Recognizing the financial barriers faced by many residents, the project offers flexible subscription options that require little to no upfront cost. Participants can choose a payment plan that suits their budget, with the assurance that their monthly energy expenses will be lower than before. This approach not only makes solar energy accessible to low-income households but also fosters a sense of ownership and investment in the community's sustainable future.

Community engagement has been a cornerstone of the initiative from the outset. Project leaders conducted extensive outreach to ensure that residents were informed and involved in every step of the process. Town hall meetings, workshops, and informational sessions provided opportunities for residents to learn about the benefits of solar energy, ask questions, and voice their concerns. This participatory approach helped build trust and enthusiasm for the project, ensuring that it was tailored to meet the community's unique needs and priorities.

The environmental benefits of the solar initiative are significant, contributing to cleaner air and a healthier environment for all. By displacing fossil fuel-based electricity with renewable solar power, the project has reduced greenhouse gas emissions and improved local air quality. These environmental gains are particularly important in low-income areas, where residents often bear the brunt of pollution and its associated health impacts. The solar initiative not only provides economic relief but also enhances the overall quality of life for the community.

The project's success has also spurred economic development, creating jobs and stimulating local businesses. From the installation and maintenance of the solar array to the administration of the subscription program, the initiative has generated employment opportunities for residents. By prioritizing local labor and resources, the project has strengthened the community's economy and provided valuable workforce training and development. These economic benefits extend beyond the immediate participants, contributing to the broader revitalization of the neighborhood.

The ripple effects of the solar initiative are being felt far beyond the boundaries of the community. Inspired by its success, neighboring areas are exploring similar projects, eager to replicate the model and reap the benefits of renewable energy. The initiative has become a catalyst for change, demonstrating the potential of solar energy to transform low-income communities and promote a more equitable and sustainable energy future.

Another compelling case study comes from a rural village nestled in the rolling hills of a developing country. Here, a solar-powered microgrid has brought electricity to a community that previously relied on kerosene lamps and diesel generators. The microgrid, powered by a combination of solar panels and battery storage, provides a reliable and affordable source of electricity for homes, schools, and businesses.

The project was initiated by a partnership between an international non-profit organization and a local community cooperative. Together, they worked to design and implement a solar solution that met the village's specific needs and constraints. The cooperative model ensured that the community had a stake in the project's success, with members contributing to the planning, financing, and operation of the microgrid.

The impact of the solar microgrid has been transformative. Access to electricity has improved the quality of life for residents, enabling them to power lights, refrigerators, and communication devices. Schools now have the resources to offer evening classes, expanding educational opportunities for children and adults alike. Local businesses have flourished, with reliable electricity supporting everything from small shops to agricultural processing facilities.

The environmental benefits of the microgrid are also noteworthy. By replacing diesel generators with clean solar power, the project has significantly reduced carbon emissions and air pollution. This shift has had positive health impacts for residents, who no longer suffer from the harmful effects of indoor air pollution caused by burning kerosene and diesel.

The success of the solar microgrid has inspired other villages in the region to pursue similar projects. By sharing their experiences and lessons learned, the community has become a leader in the movement for renewable energy access in rural areas. The project serves as a powerful example of how solar energy can drive sustainable development and improve the lives of people in underserved regions.

These case studies illustrate the transformative potential of solar energy initiatives in low-income areas. By addressing financial barriers, engaging communities, and leveraging innovative technologies, these projects have demonstrated that renewable energy can be a powerful tool for social and economic empowerment. As more communities embrace solar solutions, the vision of a sustainable and equitable energy future becomes increasingly attainable.

Chapter 2: Wind Energy: Harnessing Nature's Gift for All

The Mechanics of Wind Power and Its Accessibility

Wind power, a cornerstone of renewable energy, harnesses the kinetic energy of wind to generate electricity. This process, both elegant and efficient, involves a series of mechanical and electrical components working in harmony. Understanding the mechanics of wind power and its accessibility is crucial for appreciating its role in the global energy landscape and its potential to drive sustainable development.

At the heart of wind power generation are wind turbines, towering structures that capture the wind's energy. These turbines consist of several key components: the rotor blades, the nacelle, and the tower. The rotor blades, typically three in number, are designed to catch the wind and rotate. Their aerodynamic shape allows them to convert the wind's kinetic energy into mechanical energy. The nacelle, perched atop the tower, houses the gearbox and generator. As the rotor blades turn, they spin a shaft connected to the gearbox, which increases the rotational speed and drives the generator. The generator then converts the mechanical energy into electrical energy, which can be fed into the power grid.

The efficiency of a wind turbine depends on several factors, including the design of the blades, the height of the tower, and the wind speed at the site. Taller towers can access stronger and more consistent winds, while advanced blade designs can capture more energy from the wind. The location of a wind farm is critical, as areas with high average wind speeds are more suitable for wind power generation. Coastal regions, open plains, and hilltops are often ideal sites for wind farms due to their favorable wind conditions.

Wind power's accessibility has improved significantly over the years, driven by technological advancements and cost reductions. The cost of wind energy has decreased dramatically, making it one of the most competitive sources of electricity. This affordability has spurred the development of wind farms worldwide, from large-scale offshore installations to small community projects. As a result, wind power has become an increasingly viable option for both developed and developing countries seeking to diversify their energy portfolios and reduce their carbon footprints.

One of the key factors contributing to the accessibility of wind power is the modular nature of wind turbines. Unlike traditional power plants, which require significant infrastructure and investment, wind farms can be scaled to meet the needs of different communities. Small wind turbines can be installed in rural or remote areas, providing electricity to off-grid communities and reducing reliance on diesel generators. These small-scale installations are particularly beneficial in developing regions, where access to electricity is limited and the cost of extending the grid is prohibitive.

Community wind projects have emerged as a powerful tool for promoting local energy independence and economic development. By investing in wind power, communities can generate their own electricity, reduce energy costs, and create jobs. These projects often involve partnerships between local governments, non-profit organizations, and private companies, ensuring that the benefits of wind power are shared equitably.

Community wind projects also foster a sense of ownership and pride, as residents take an active role in shaping their energy future.

The environmental benefits of wind power are significant, contributing to cleaner air and a healthier planet. Wind energy is a clean, renewable resource that produces no greenhouse gas emissions during operation. By displacing fossil fuel-based electricity, wind power helps mitigate climate change and reduce air pollution. This shift has positive health impacts, particularly in areas where air quality is a concern. Additionally, wind farms have a relatively small land footprint, allowing for multiple land uses, such as agriculture or conservation, to coexist with energy production.

Despite its many advantages, wind power faces several challenges that must be addressed to ensure its continued growth and accessibility. One of the primary concerns is the intermittency of wind, as wind speeds can vary significantly over time. This variability can affect the stability and reliability of the power grid, particularly in regions with high wind penetration. To address this issue, grid operators are investing in advanced forecasting techniques and energy storage solutions, such as batteries and pumped hydro storage, to balance supply and demand.

Another challenge is the potential impact of wind farms on wildlife, particularly birds and bats. While the overall impact is relatively low compared to other human activities, it is essential to minimize any negative effects through careful site selection and mitigation measures. Researchers are exploring innovative technologies, such as radar systems and acoustic deterrents, to reduce wildlife collisions and enhance the environmental sustainability of wind power.

Public perception and acceptance of wind power can also influence its accessibility. While many people support renewable energy, concerns about noise, visual impact, and land use can lead to opposition from local communities. Engaging with stakeholders early in the planning process and addressing their concerns through transparent communication and community involvement is crucial for building trust and support for wind projects.

The future of wind power is promising, with significant potential for growth and innovation. Technological advancements, such as larger and more efficient turbines, floating offshore wind farms, and hybrid systems that combine wind with other renewable sources, are expected to drive further cost reductions and expand the reach of wind energy. By continuing to invest in research and development, policymakers and industry leaders can ensure that wind power remains a key component of the global energy transition.

Wind power's accessibility is also enhanced by supportive policies and regulatory frameworks. Government incentives, such as tax credits, feed-in tariffs, and renewable energy mandates, have been instrumental in driving the adoption of wind energy. By creating a favorable regulatory environment, policymakers can encourage investment in wind projects and facilitate the integration of wind power into the energy mix.

Education and outreach efforts are essential for raising awareness about the benefits of wind power and promoting its adoption. By providing information about the environmental and economic advantages of wind energy, individuals and communities can make informed decisions about their energy needs. Workshops, training programs, and informational resources can empower users to take advantage of the latest innovations in wind technology, fostering a culture of sustainability and environmental stewardship.

As the world continues to grapple with the challenges of climate change and energy security, wind power offers a sustainable and accessible solution. By harnessing the power of the wind, we can reduce our reliance on fossil fuels, decrease greenhouse gas emissions, and create a cleaner, healthier planet for future generations. Through continued innovation, collaboration, and commitment, wind power can play a pivotal role in shaping a sustainable energy future.

Small-Scale Wind Solutions for Rural and Remote Areas

In the quiet expanse of rural and remote areas, where the hum of urban life fades into the distance, small-scale wind solutions offer a beacon of hope and sustainability. These areas, often characterized by limited access to centralized power grids, face unique challenges in meeting their energy needs. Small-scale wind turbines, with their ability to harness the natural power of the wind, provide an effective and sustainable solution for these communities, transforming the way they generate and consume energy.

The appeal of small-scale wind solutions lies in their adaptability and efficiency. Unlike large wind farms that require significant infrastructure and investment, small-scale turbines can be tailored to meet the specific needs of individual households, farms, or small businesses. These turbines, typically ranging from a few hundred watts to several kilowatts in capacity, are designed to operate efficiently in a variety of wind conditions, making them ideal for rural and remote areas where wind patterns can be unpredictable.

One of the primary advantages of small-scale wind solutions is their ability to provide energy independence. For communities located far from the nearest power grid, the cost and logistics of extending electrical lines can be prohibitive. Small-scale wind turbines offer a decentralized energy source, allowing residents to generate their own electricity and reduce their reliance on external power supplies. This independence not only enhances energy security but also empowers communities to take control of their energy future.

The installation of small-scale wind turbines can also lead to significant cost savings over time. While the initial investment may be a consideration, the long-term benefits often outweigh the upfront costs. By generating their own electricity, residents can reduce or eliminate their monthly energy bills, freeing up resources for other essential needs. Additionally, many governments offer incentives and rebates for renewable energy installations, further offsetting the initial expenses and making small-scale wind solutions more accessible.

The environmental benefits of small-scale wind solutions are equally compelling. Wind energy is a clean, renewable resource that produces no greenhouse gas emissions during operation. By replacing or supplementing fossil fuel-based energy sources, small-scale wind turbines contribute to a reduction in carbon emissions and air pollution. This shift is particularly important in rural and remote areas, where residents may rely on diesel generators or other polluting energy sources. The adoption of wind energy can lead to improved air quality and a healthier environment for all.

The versatility of small-scale wind solutions extends beyond electricity generation. In agricultural settings, wind turbines can be used to power irrigation systems, water pumps, and other essential equipment. This capability is particularly valuable in regions where water resources are scarce or difficult to access. By harnessing the power of the wind, farmers can improve their productivity and resilience, ensuring a stable food supply for their communities.

Community engagement and education are critical components of successful small-scale wind projects. By involving residents in the planning and implementation process, project leaders can ensure that the solutions are tailored to meet the specific needs and preferences of the community. Workshops, training sessions, and informational resources can empower residents to understand the benefits of wind energy and how to maintain and operate their turbines effectively. This participatory approach fosters a sense of ownership and pride, as communities work together to achieve their energy goals.

The success of small-scale wind solutions in rural and remote areas is exemplified by numerous case studies around the world. In a remote village nestled in the mountains, a community wind project has transformed the lives of its residents. Prior to the installation of small-scale wind turbines, the village relied on kerosene lamps and diesel generators for lighting and power. The introduction of wind energy has provided a reliable and affordable source of electricity, enabling residents to power lights, refrigerators, and communication devices. The project has also created jobs and stimulated local economic development, as residents have been trained to install and maintain the turbines.

In another example, a farming community in a windswept region has embraced small-scale wind solutions to power their irrigation systems. The turbines, strategically placed throughout the fields, provide the energy needed to pump water from underground wells, ensuring a consistent water supply for crops. This innovation has increased agricultural productivity and resilience, allowing farmers to expand their operations and improve their livelihoods.

The potential for small-scale wind solutions to drive sustainable development in rural and remote areas is immense. By providing a reliable and affordable source of energy, these solutions can improve the quality of life for residents, enhance economic opportunities, and promote environmental stewardship. As technology continues to advance and costs decline, the accessibility and appeal of small-scale wind solutions are expected to grow, making them an increasingly viable option for communities around the world.

Challenges remain, however, in the widespread adoption of small-scale wind solutions. One of the primary obstacles is the variability of wind resources, which can affect the performance and reliability of wind turbines. To address this issue, careful site assessment and planning are essential to ensure that turbines are installed in locations with adequate wind speeds. Additionally, advancements in turbine design and technology are helping to improve efficiency and performance in low-wind conditions.

Another challenge is the need for ongoing maintenance and support to ensure the long-term success of small-scale wind projects. While wind turbines are generally low-maintenance, regular inspections and servicing are necessary to keep them operating at peak efficiency. Training programs and technical support services can help communities develop the skills and knowledge needed to maintain their turbines and address any issues that arise.

The integration of small-scale wind solutions with other renewable energy sources, such as solar power, can further enhance their effectiveness and reliability. Hybrid systems that combine wind and solar energy can provide a more consistent and stable power supply, particularly in regions where wind and solar resources complement each other. By diversifying their energy sources, communities can increase their resilience and reduce their vulnerability to fluctuations in weather patterns.

The future of small-scale wind solutions in rural and remote areas is bright, with significant potential for growth and innovation. As more communities embrace the benefits of wind energy, the vision of a sustainable and equitable energy future becomes increasingly attainable. Through continued collaboration, investment, and commitment, small-scale wind solutions can play a pivotal role in transforming the energy landscape and improving the lives of people around the world.

Policy and Incentives for Community Wind Projects

Community wind projects have emerged as a vital component of the renewable energy landscape, offering a pathway to sustainable energy generation that is both locally controlled and economically beneficial. These projects, which involve the development of wind energy resources by and for local communities, rely heavily on supportive policies and incentives to thrive. Understanding the policy landscape and available incentives is crucial for communities seeking to embark on their own wind energy journey.

The foundation of successful community wind projects often lies in the regulatory frameworks established by governments at various levels. These frameworks can either facilitate or hinder the development of wind energy, depending on how they are structured. Key policy elements that support community wind projects include streamlined permitting processes, favorable zoning regulations, and clear guidelines for grid interconnection. By reducing bureaucratic hurdles and providing a clear path for project development, these policies can significantly lower the barriers to entry for community wind initiatives.

Financial incentives play a pivotal role in making community wind projects economically viable. One of the most common forms of support is the provision of tax credits, which can offset the costs of wind turbine installation and operation. For example, production tax credits (PTCs) and investment tax credits (ITCs) have been instrumental in driving the growth of wind energy in many regions. These credits provide financial relief by reducing the tax liability of project developers, thereby improving the overall return on investment.

Grants and subsidies are another important tool for supporting community wind projects. These funds can be used to cover a portion of the capital costs associated with wind turbine installation, making it easier for communities to finance their projects. Government agencies, non-profit organizations, and private foundations often offer grant programs specifically targeted at renewable energy initiatives. By securing grant funding, communities can reduce their reliance on debt financing and lower the financial risk associated with their projects.

Feed-in tariffs (FITs) represent a policy mechanism that guarantees a fixed payment for the electricity generated by wind projects over a specified period. This approach provides long-term revenue certainty for project developers, encouraging investment in wind energy. FITs can be particularly beneficial for community wind projects, as they offer a stable income stream that can be used to repay loans, cover operational costs, and reinvest in community development initiatives.

Net metering policies also support community wind projects by allowing project participants to receive credit for the excess electricity they generate and feed back into the grid. This arrangement effectively reduces the participants' electricity bills, enhancing the economic attractiveness of wind energy. Net metering can be especially advantageous for community wind projects that involve multiple stakeholders, as it ensures that all participants benefit from the energy produced.

Community ownership models are a defining feature of many successful wind projects, fostering local engagement and ensuring that the economic benefits of wind energy are retained within the community. Policies that support community ownership, such as cooperative structures and shared ownership models, can empower residents to take an active role in the development and operation of wind projects. By pooling resources and sharing risks, community ownership models can make wind energy more accessible and inclusive.

Public-private partnerships (PPPs) offer a collaborative approach to developing community wind projects, leveraging the strengths of both sectors to achieve common goals. Through PPPs, communities can access the technical expertise, financial resources, and project management capabilities of private companies, while retaining control over project outcomes. Policies that encourage and facilitate PPPs can enhance the feasibility and success of community wind initiatives.

Education and outreach efforts are essential for building public support and understanding of community wind projects. By providing information about the benefits of wind energy, the project development process, and available incentives, communities can foster a culture of sustainability and environmental stewardship. Workshops, informational sessions, and community meetings can engage residents and stakeholders, ensuring that projects are aligned with local needs and priorities.

The role of local governments in supporting community wind projects cannot be overstated. By adopting policies that prioritize renewable energy development, local governments can create an enabling environment for wind projects to flourish. This may include setting renewable energy targets, offering financial incentives, and facilitating community engagement. Local governments can also serve as project champions, advocating for wind energy and building partnerships with other stakeholders.

Case studies of successful community wind projects offer valuable insights and lessons for other communities seeking to embark on similar initiatives. In one example, a rural community formed a cooperative to develop a wind farm, leveraging government grants and tax credits to finance the project. The cooperative model allowed residents to pool their resources and share in the project's profits, which were reinvested in local infrastructure and services. This approach not only provided a reliable source of clean energy but also stimulated economic development and strengthened community ties.

In another instance, an urban neighborhood partnered with a private company to install wind turbines on a nearby industrial site. The project was supported by a combination of feed-in tariffs and net metering policies, ensuring a stable revenue stream for both the community and the company. The partnership enabled the neighborhood to benefit from reduced energy costs and increased energy independence, while the company gained access to a new market and enhanced its corporate social responsibility profile.

The future of community wind projects is promising, with significant potential for growth and innovation. As technology continues to advance and costs decline, wind energy is becoming an increasingly attractive option for communities seeking to transition to renewable energy. By leveraging supportive policies and incentives, communities can overcome financial and regulatory barriers, ensuring that the benefits of wind energy are accessible to all.

The integration of community wind projects with other renewable energy sources, such as solar power, can further enhance their effectiveness and resilience. Hybrid systems that combine wind and solar energy can provide a more consistent and stable power supply, particularly in regions where wind and solar resources complement each other. By diversifying their energy sources, communities can increase their resilience and reduce their vulnerability to fluctuations in weather patterns.

As the world continues to grapple with the challenges of climate change and energy security, community wind projects offer a sustainable and equitable solution. By harnessing the power of the wind, communities can reduce their reliance on fossil fuels, decrease greenhouse gas emissions, and create a cleaner, healthier planet for future generations. Through continued innovation, collaboration, and commitment, community wind projects can play a pivotal role in shaping a sustainable energy future.

Addressing Social and Environmental Concerns in Wind Development

Wind energy, a cornerstone of the renewable revolution, offers a promising path toward a sustainable future. However, as with any large-scale development, it brings with it a set of social and environmental concerns that must be addressed to ensure its success and acceptance. Understanding these concerns and implementing strategies to mitigate them is crucial for the harmonious integration of wind projects into communities and ecosystems.

One of the primary social concerns associated with wind development is the potential impact on local communities. Wind farms, with their towering turbines, can alter the landscape and affect the visual aesthetics of an area. For residents accustomed to unobstructed views of natural scenery, the introduction of wind turbines can be jarring. To address this, developers must engage with communities early in the planning process, providing clear and transparent information about the project and its benefits. Visual simulations and site visits can help residents visualize the changes and understand the project's scope.

Noise is another concern often raised by communities near wind farms. The mechanical sounds generated by turbines, though generally low, can be a source of annoyance for some residents, particularly in quiet rural areas. To mitigate noise impacts, developers can employ advanced turbine designs that minimize sound emissions and strategically site turbines to maintain a buffer distance from residential areas. Regular monitoring and maintenance can also ensure that turbines operate quietly and efficiently.

The potential impact on property values is a concern for many homeowners living near proposed wind farms. While studies have shown mixed results, with some indicating minimal impact and others suggesting a decrease in property values, the perception of potential loss can fuel opposition. Developers can address these concerns by demonstrating the economic benefits of wind projects, such as job creation, increased tax revenue, and community investment. By highlighting the positive contributions of wind energy, developers can build support and alleviate fears about property values.

Community engagement is a critical component of addressing social concerns in wind development. By involving residents in the decision-making process, developers can foster a sense of ownership and collaboration. Community advisory panels, public meetings, and open forums provide opportunities for residents to voice their concerns, ask questions, and offer input. This participatory approach not only builds trust but also ensures that projects are tailored to meet the needs and preferences of the community.

Environmental concerns are equally important in the development of wind energy projects. One of the most significant issues is the potential impact on wildlife, particularly birds and bats. Wind turbines can pose a collision risk for these animals, leading to mortality and population declines. To mitigate this impact, developers can conduct thorough environmental assessments to identify sensitive habitats and migration routes. By selecting sites with minimal wildlife activity and implementing mitigation measures, such as turbine curtailment during peak migration periods, developers can reduce the risk to wildlife.

The construction and operation of wind farms can also affect local ecosystems and land use. The development process may involve clearing vegetation, altering drainage patterns, and disrupting soil. To minimize these impacts, developers can employ best practices for site preparation and restoration, such as replanting native vegetation and implementing erosion control measures. Additionally, wind farms can be designed to coexist with other land uses, such as agriculture or conservation, allowing for multiple benefits from the same land area.

The decommissioning of wind turbines at the end of their operational life is another environmental consideration. Proper planning for decommissioning ensures that turbines are dismantled safely and responsibly, with materials recycled or disposed of appropriately. Developers can include decommissioning plans as part of the initial project proposal, outlining the steps and responsibilities involved in the process. By planning for the entire lifecycle of a wind project, developers can ensure that environmental impacts are minimized from start to finish.

Cultural and historical preservation is an important aspect of wind development, particularly in areas with significant heritage sites. Wind projects must be carefully planned to avoid disrupting or damaging cultural landmarks and archaeological sites. Collaboration with local cultural and historical organizations can help identify and protect these sites, ensuring that wind development respects and preserves the cultural heritage of the area.

The integration of wind energy into the existing power grid presents both opportunities and challenges. While wind energy can contribute to a more sustainable and resilient energy system, its variability can pose challenges for grid stability and reliability. To address this, grid operators can invest in advanced forecasting techniques, energy storage solutions, and grid infrastructure upgrades. By enhancing the flexibility and capacity of the grid, operators can accommodate the intermittent nature of wind energy and ensure a reliable power supply.

Public perception and acceptance of wind energy are influenced by a variety of factors, including personal values, experiences, and information sources. To build public support, developers and policymakers can engage in proactive communication and education efforts. By providing accurate and accessible information about the benefits and challenges of wind energy, stakeholders can foster a more informed and supportive public. Educational campaigns, school programs, and community workshops can raise awareness and promote a positive image of wind energy.

The economic benefits of wind development can be a powerful tool for addressing social concerns and building support. Wind projects can create jobs, stimulate local economies, and generate tax revenue for communities. By highlighting these benefits and ensuring that they are distributed equitably, developers can demonstrate the value of wind energy to local residents.

Community benefit agreements, which outline the economic contributions of a project, can provide a formal mechanism for ensuring that communities receive a fair share of the benefits.

The future of wind development depends on the ability to address social and environmental concerns effectively. By adopting a holistic approach that considers the needs and values of communities and ecosystems, developers can create projects that are both sustainable and socially responsible. Through collaboration, innovation, and commitment, wind energy can continue to play a vital role in the transition to a cleaner and more sustainable energy future.

Future Prospects for Inclusive Wind Energy Solutions

The horizon of wind energy is expanding, promising a future where inclusive solutions can power communities across the globe. As the world grapples with the dual challenges of climate change and energy access, wind energy stands out as a beacon of hope, offering sustainable and equitable solutions. The future of wind energy is not just about technological advancements but also about ensuring that its benefits are accessible to all, regardless of geographic, economic, or social barriers.

The evolution of wind technology is paving the way for more inclusive energy solutions. Innovations in turbine design, materials, and efficiency are making wind energy more affordable and adaptable. Smaller, more efficient turbines are being developed to cater to diverse environments, from urban rooftops to remote rural areas. These advancements are crucial for extending the reach of wind energy to communities that have traditionally been underserved by conventional energy systems.

Offshore wind energy is emerging as a significant player in the future energy landscape. With vast untapped potential, offshore wind farms can generate substantial amounts of clean energy without competing for land resources. Floating turbine technology is particularly promising, allowing for the deployment of wind farms in deeper waters where wind speeds are higher and more consistent. This development opens up new possibilities for coastal and island communities, providing them with a reliable and sustainable energy source.

The integration of wind energy with other renewable sources is another avenue for creating inclusive solutions. Hybrid systems that combine wind with solar, hydro, or biomass energy can provide a more stable and reliable power supply, particularly in regions with variable weather patterns. By diversifying energy sources, communities can enhance their resilience and reduce their dependence on fossil fuels. This approach not only ensures a consistent energy supply but also maximizes the use of local resources, fostering energy independence.

Energy storage technologies are playing a pivotal role in the future of wind energy. As wind is an intermittent resource, the ability to store excess energy for use during periods of low wind is essential for maintaining a stable power supply. Advances in battery technology, such as lithium-ion and flow batteries, are making energy storage more efficient and cost-effective. These technologies enable communities to harness the full potential of wind energy, ensuring that it is available whenever needed.

Policy frameworks and incentives are critical for promoting inclusive wind energy solutions. Governments and regulatory bodies can create an enabling environment by implementing policies that support renewable energy development, such as feed-in tariffs, tax credits, and grants. These incentives can lower the financial barriers to entry for communities and encourage investment in wind projects. By prioritizing policies that promote equity and access, policymakers can ensure that the benefits of wind energy are shared widely.

Community engagement and participation are essential components of inclusive wind energy solutions. By involving local residents in the planning and decision-making process, developers can ensure that projects are aligned with community needs and values. This participatory approach fosters a sense of ownership and empowerment, as communities take an active role in shaping their energy future. Community wind projects, where residents collectively own and benefit from wind farms, exemplify this model of inclusivity and collaboration.

Education and capacity-building initiatives are vital for empowering communities to embrace wind energy. By providing training and resources, communities can develop the skills and knowledge needed to implement and maintain wind projects. Educational programs can raise awareness about the benefits of wind energy and inspire the next generation of renewable energy leaders. By investing in education, communities can build a foundation for long-term sustainability and innovation.

The economic benefits of wind energy extend beyond energy generation, offering opportunities for job creation and economic development. The construction, operation, and maintenance of wind farms require a skilled workforce, creating jobs in engineering, manufacturing, and technical services. By investing in local workforce development, communities can ensure that the economic benefits of wind energy are retained locally, contributing to economic resilience and prosperity.

The environmental benefits of wind energy are well-documented, but its role in promoting social equity is equally important. By providing a clean and affordable energy source, wind energy can improve the quality of life for underserved communities, reducing energy poverty and enhancing access to essential services. This shift has positive implications for health, education, and economic opportunities, contributing to a more equitable and just society.

The future of wind energy is not without challenges, but the potential for inclusive solutions is immense. By addressing barriers to access and ensuring that the benefits of wind energy are distributed equitably, we can create a more sustainable and inclusive energy future. Through innovation, collaboration, and commitment, wind energy can play a transformative role in addressing the global energy challenge and building a better world for all.

As we look to the future, the vision of inclusive wind energy solutions is within reach. By harnessing the power of the wind, we can create a cleaner, healthier, and more equitable planet for future generations. The journey toward this vision requires dedication, creativity, and a shared commitment to sustainability and justice. Together, we can unlock the full potential of wind energy and pave the way for a brighter, more inclusive future.

Chapter 3: Hydropower: Making Waves in Energy Accessibility

Understanding Hydropower's Role in Renewable Energy Access

Hydropower, a cornerstone of renewable energy, has long played a pivotal role in providing clean and reliable electricity. As the world seeks to transition to sustainable energy sources, understanding hydropower's contribution to renewable energy access is essential. This chapter delves into the multifaceted role of hydropower, exploring its benefits, challenges, and potential for expanding energy access globally.

Hydropower harnesses the energy of flowing or falling water to generate electricity, making it one of the oldest and most established forms of renewable energy. Its appeal lies in its ability to provide a consistent and reliable power supply, unlike some other renewable sources that are subject to weather variability. This reliability makes hydropower a crucial component of the energy mix, particularly in regions where energy demand is high and consistent supply is essential.

One of the primary advantages of hydropower is its capacity for large-scale electricity generation. Large hydropower plants, often built on major rivers, can produce significant amounts of electricity, contributing to national grids and supporting industrial and urban development. These projects can provide a stable and continuous power supply, reducing reliance on fossil fuels and lowering greenhouse gas emissions. In countries with abundant water resources, hydropower can serve as the backbone of the energy system, supporting economic growth and development.

Beyond large-scale projects, small and micro hydropower systems offer opportunities for decentralized energy generation, particularly in rural and remote areas. These systems, which can be installed on small rivers or streams, provide a sustainable energy solution for communities that are not connected to the national grid. By generating electricity locally, small hydropower projects can enhance energy access, improve living standards, and stimulate local economic development. This decentralized approach empowers communities to take control of their energy future, reducing dependence on external energy sources.

Hydropower's role in renewable energy access extends beyond electricity generation. Pumped storage hydropower, a type of energy storage system, plays a critical role in balancing supply and demand on the grid. By storing excess electricity during periods of low demand and releasing it during peak demand, pumped storage systems enhance grid stability and reliability. This capability is particularly valuable in integrating variable renewable energy sources, such as wind and solar, into the grid. By providing a flexible and responsive energy storage solution, hydropower supports the transition to a more sustainable and resilient energy system.

The environmental benefits of hydropower are significant, but they must be balanced with potential ecological and social impacts. Large hydropower projects can alter river ecosystems, affecting fish populations, water quality, and sediment transport. To mitigate these impacts, developers must conduct thorough environmental assessments and implement measures to protect and restore ecosystems. Fish ladders, bypass channels, and habitat restoration are examples of strategies that can minimize the ecological footprint of hydropower projects.

Social considerations are equally important in hydropower development. Large projects can lead to the displacement of communities and changes in land use, raising concerns about social equity and justice. Engaging with affected communities and ensuring their participation in decision-making processes is crucial for addressing these concerns. Compensation, resettlement, and benefit-sharing mechanisms can help ensure that communities are not only compensated for their losses but also benefit from the development of hydropower projects.

The potential for expanding hydropower's role in renewable energy access is immense, but it requires careful planning and consideration of local contexts. In regions with untapped water resources, there is significant potential for developing new hydropower projects. However, these projects must be designed to minimize environmental and social impacts, ensuring that they contribute to sustainable development goals. By adopting best practices and innovative technologies, developers can enhance the sustainability and inclusivity of hydropower projects.

Technological advancements are driving the evolution of hydropower, making it more efficient and adaptable. Innovations in turbine design, materials, and control systems are improving the efficiency and performance of hydropower plants. These advancements enable the development of projects in a wider range of environments, including low-head and variable-flow sites. By expanding the range of viable sites, technology is opening up new opportunities for hydropower development, particularly in regions with limited traditional resources.

Policy frameworks and incentives play a crucial role in promoting hydropower development and ensuring its contribution to renewable energy access. Governments can create an enabling environment by implementing policies that support sustainable hydropower development, such as streamlined permitting processes, financial incentives, and environmental regulations. By prioritizing policies that promote equity and access, policymakers can ensure that the benefits of hydropower are shared widely and contribute to broader development goals.

Community engagement and participation are essential components of successful hydropower projects. By involving local residents in the planning and decision-making process, developers can ensure that projects are aligned with community needs and values. This participatory approach fosters a sense of ownership and empowerment, as communities take an active role in shaping their energy future. Community-based hydropower projects, where residents collectively own and benefit from the energy generated, exemplify this model of inclusivity and collaboration.

Education and capacity-building initiatives are vital for empowering communities to embrace hydropower. By providing training and resources, communities can develop the skills and knowledge needed to implement and maintain hydropower projects. Educational programs can raise awareness about the benefits of hydropower and inspire the next generation of renewable energy leaders. By investing in education, communities can build a foundation for long-term sustainability and innovation.

The economic benefits of hydropower extend beyond energy generation, offering opportunities for job creation and economic development. The construction, operation, and maintenance of hydropower plants require a skilled workforce, creating jobs in engineering, manufacturing, and technical services. By investing in local workforce development, communities can ensure that the economic benefits of hydropower are retained locally, contributing to economic resilience and prosperity.

The future of hydropower is promising, with significant potential for growth and innovation. As technology continues to advance and costs decline, hydropower is becoming an increasingly attractive option for communities seeking to transition to renewable energy. By leveraging supportive policies and incentives, communities can overcome financial and regulatory barriers, ensuring that the benefits of hydropower are accessible to all.

As the world continues to grapple with the challenges of climate change and energy security, hydropower offers a sustainable and equitable solution. By harnessing the power of water, communities can reduce their reliance on fossil fuels, decrease greenhouse gas emissions, and create a cleaner, healthier planet for future generations. Through continued innovation, collaboration, and commitment, hydropower can play a pivotal role in shaping a sustainable energy future.

Micro-Hydropower Systems for Local Communities

Micro-hydropower systems present a compelling opportunity for local communities to harness renewable energy in a sustainable and accessible manner. These systems, which typically generate up to 100 kilowatts of electricity, are designed to utilize the natural flow of water in small streams or rivers. By tapping into this resource, communities can achieve energy independence, reduce their carbon footprint, and stimulate local economic development.

The appeal of micro-hydropower lies in its simplicity and adaptability. Unlike large-scale hydropower projects, which often require significant infrastructure and investment, micro-hydropower systems can be installed with minimal environmental disruption. This makes them particularly suitable for rural and remote areas where access to the national grid is limited or non-existent. By providing a reliable source of electricity, micro-hydropower can transform the lives of residents, powering homes, schools, and businesses.

One of the key advantages of micro-hydropower is its ability to operate continuously, providing a stable and consistent energy supply. Unlike solar or wind energy, which are subject to weather variability, micro-hydropower systems can generate electricity day and night, as long as there is a sufficient flow of water. This reliability is crucial for communities that rely on electricity for essential services, such as healthcare and education.

The process of developing a micro-hydropower system begins with a thorough assessment of the local water resources. Identifying a suitable site is critical, as the system's efficiency depends on the flow rate and head (the vertical distance the water falls). Once a site is selected, a feasibility study is conducted to evaluate the technical and economic viability of the project. This study considers factors such as water availability, environmental impact, and potential energy output.

Community involvement is a vital component of successful micro-hydropower projects. By engaging local residents in the planning and decision-making process, developers can ensure that the project meets the community's needs and priorities. This participatory approach fosters a sense of ownership and empowerment, as residents take an active role in shaping their energy future. Community meetings, workshops, and training sessions can provide valuable opportunities for residents to learn about the benefits of micro-hydropower and contribute their insights and ideas.

The installation of a micro-hydropower system involves several key components, including a water intake, penstock, turbine, generator, and control system. The water intake diverts a portion of the stream's flow into the penstock, a pipe that channels the water to the turbine. As the water flows through the turbine, it spins the blades, converting the kinetic energy of the water into mechanical energy. This mechanical energy is then transformed into electrical energy by the generator. The control system regulates the flow of electricity, ensuring that it is distributed efficiently and safely to the community.

Environmental considerations are an important aspect of micro-hydropower development. While these systems have a relatively low environmental impact compared to large-scale projects, they can still affect local ecosystems. To minimize these impacts, developers can implement measures such as fish screens, bypass channels, and habitat restoration. These strategies help protect aquatic life and maintain the ecological balance of the stream or river.

The economic benefits of micro-hydropower extend beyond energy generation. By providing a reliable source of electricity, these systems can support local businesses, create jobs, and stimulate economic growth. For example, access to electricity can enable the establishment of small enterprises, such as milling, refrigeration, and telecommunications, which can enhance the community's economic resilience and prosperity. Additionally, the construction, operation, and maintenance of micro-hydropower systems require skilled labor, creating employment opportunities for local residents.

Financing is a critical consideration for micro-hydropower projects, particularly in resource-constrained communities. Various funding options are available, including government grants, loans, and private investment. Community-based financing models, such as cooperatives or crowdfunding, can also be effective in mobilizing resources and ensuring that the benefits of the project are shared equitably. By pooling resources and sharing risks, communities can overcome financial barriers and achieve energy independence.

Education and capacity-building initiatives are essential for the long-term success of micro-hydropower projects. By providing training and resources, communities can develop the skills and knowledge needed to operate and maintain their systems. Educational programs can raise awareness about the benefits of renewable energy and inspire the next generation of energy leaders. By investing in education, communities can build a foundation for sustainable development and innovation.

The integration of micro-hydropower with other renewable energy sources can enhance the resilience and reliability of local energy systems. Hybrid systems that combine micro-hydropower with solar or wind energy can provide a more stable and consistent power supply, particularly in regions with variable weather patterns. By diversifying their energy sources, communities can reduce their dependence on fossil fuels and increase their energy security.

The future of micro-hydropower is promising, with significant potential for growth and innovation. As technology continues to advance and costs decline, micro-hydropower is becoming an increasingly attractive option for communities seeking to transition to renewable energy. By leveraging supportive policies and incentives, communities can overcome financial and regulatory barriers, ensuring that the benefits of micro-hydropower are accessible to all.

As the world continues to grapple with the challenges of climate change and energy security, micro-hydropower offers a sustainable and equitable solution. By harnessing the power of water, communities can reduce their reliance on fossil fuels, decrease greenhouse gas emissions, and create a cleaner, healthier planet for future generations. Through continued innovation, collaboration, and commitment, micro-hydropower can play a pivotal role in shaping a sustainable energy future.

Environmental and Social Considerations in Hydropower Projects

Hydropower projects, while offering significant benefits in terms of renewable energy generation, necessitate careful consideration of both environmental and social factors. These projects, whether large or small, can have profound impacts on ecosystems and communities. Understanding these impacts and implementing strategies to mitigate them is essential for the sustainable development of hydropower.

The environmental considerations of hydropower projects are multifaceted. One of the primary concerns is the alteration of river ecosystems. Dams and reservoirs can disrupt the natural flow of rivers, affecting water quality, sediment transport, and aquatic habitats. Fish populations, in particular, can be adversely impacted, as dams can obstruct migration routes and alter spawning grounds. To address these issues, developers can implement measures such as fish ladders, bypass channels, and habitat restoration. These strategies help maintain ecological balance and protect biodiversity.

Water quality is another critical environmental consideration. The construction and operation of hydropower projects can lead to changes in water temperature, oxygen levels, and nutrient concentrations. These changes can affect aquatic life and downstream water users. To mitigate these impacts, developers can employ techniques such as selective water withdrawal, aeration, and sediment management. By carefully managing water quality, hydropower projects can minimize their environmental footprint and support healthy ecosystems.

The creation of reservoirs for hydropower projects can also lead to the inundation of land, resulting in the loss of terrestrial habitats and biodiversity. This impact can be particularly significant in areas with high conservation value or endemic species. To minimize habitat loss, developers can conduct thorough environmental assessments to identify sensitive areas and design projects that avoid or minimize inundation. Additionally, habitat compensation and restoration programs can help offset the impacts of land loss, ensuring that biodiversity is preserved.

Social considerations are equally important in the development of hydropower projects. One of the most significant social impacts is the displacement of communities. Large hydropower projects can require the relocation of residents, leading to the loss of homes, livelihoods, and cultural heritage. To address these concerns, developers must engage with affected communities early in the planning process, ensuring that their voices are heard and their rights are respected. Compensation, resettlement, and benefit-sharing mechanisms can help ensure that displaced communities are not only compensated for their losses but also benefit from the development of hydropower projects.

The cultural and historical significance of river systems must also be considered in hydropower development. Rivers often hold deep cultural and spiritual meaning for local communities, and their alteration can have profound social and cultural impacts. Developers can work with local cultural and historical organizations to identify and protect significant sites, ensuring that hydropower projects respect and preserve cultural heritage. By incorporating cultural considerations into project planning, developers can foster positive relationships with communities and enhance the social acceptance of hydropower projects.

Community engagement and participation are essential components of successful hydropower projects. By involving local residents in the planning and decision-making process, developers can ensure that projects are aligned with community needs and values. This participatory approach fosters a sense of ownership and empowerment, as communities take an active role in shaping their energy future. Community meetings, workshops, and training sessions provide valuable opportunities for residents to learn about the benefits of hydropower and contribute their insights and ideas.

The economic benefits of hydropower projects can be significant, but they must be distributed equitably to ensure social acceptance. By providing a reliable source of electricity, hydropower projects can support local businesses, create jobs, and stimulate economic growth. For example, access to electricity can enable the establishment of small enterprises, such as milling, refrigeration, and telecommunications, which can enhance the community's economic resilience and prosperity. Additionally, the construction, operation, and maintenance of hydropower projects require skilled labor, creating employment opportunities for local residents.

Financing is a critical consideration for hydropower projects, particularly in resource-constrained communities. Various funding options are available, including government grants, loans, and private investment. Community-based financing models, such as cooperatives or crowdfunding, can also be effective in mobilizing resources and ensuring that the benefits of the project are shared equitably. By pooling resources and sharing risks, communities can overcome financial barriers and achieve energy independence.

Education and capacity-building initiatives are vital for the long-term success of hydropower projects. By providing training and resources, communities can develop the skills and knowledge needed to operate and maintain their systems. Educational programs can raise awareness about the benefits of renewable energy and inspire the next generation of energy leaders. By investing in education, communities can build a foundation for sustainable development and innovation.

The integration of hydropower with other renewable energy sources can enhance the resilience and reliability of local energy systems. Hybrid systems that combine hydropower with solar or wind energy can provide a more stable and consistent power supply, particularly in regions with variable weather patterns. By diversifying their energy sources, communities can reduce their dependence on fossil fuels and increase their energy security.

The future of hydropower is promising, with significant potential for growth and innovation. As technology continues to advance and costs decline, hydropower is becoming an increasingly attractive option for communities seeking to transition to renewable energy. By leveraging supportive policies and incentives, communities can overcome financial and regulatory barriers, ensuring that the benefits of hydropower are accessible to all.

As the world continues to grapple with the challenges of climate change and energy security, hydropower offers a sustainable and equitable solution. By harnessing the power of water, communities can reduce their reliance on fossil fuels, decrease greenhouse gas emissions, and create a cleaner, healthier planet for future generations. Through continued innovation, collaboration, and commitment, hydropower can play a pivotal role in shaping a sustainable energy future.

Case Studies: Hydropower Success Stories in Developing Regions

Hydropower has emerged as a transformative force in developing regions, offering sustainable energy solutions that drive economic growth and improve quality of life. Across the globe, numerous communities have harnessed the power of water to generate electricity, demonstrating the potential of hydropower to address energy poverty and foster development.

This chapter delves into several case studies that highlight the success of hydropower projects in diverse settings, showcasing the innovative approaches and collaborative efforts that have led to their achievements.

In the heart of East Africa, Rwanda's Nyabarongo Hydropower Project stands as a testament to the power of collaboration and strategic planning. Commissioned in 2015, this project was developed to address the country's pressing energy needs and reduce its reliance on imported fossil fuels. Situated on the Nyabarongo River, the plant has a capacity of 28 megawatts, providing a significant boost to Rwanda's national grid. The project's success can be attributed to a combination of factors, including strong government support, international partnerships, and a commitment to environmental sustainability. By prioritizing local workforce development and community engagement, the Nyabarongo project has not only enhanced energy access but also created jobs and stimulated economic growth in the region.

In South Asia, Nepal's Chilime Hydropower Plant serves as a model for community-based energy development. Located in the Rasuwa District, this 22-megawatt project was initiated by the Nepal Electricity Authority in collaboration with local communities. The plant's innovative financing model, which involved local equity participation, ensured that the benefits of the project were shared equitably among stakeholders. By involving local residents in the planning and decision-making process, the Chilime project fostered a sense of ownership and empowerment, leading to its successful implementation. The plant has not only improved energy access in the region but also generated revenue for community development initiatives, such as education and healthcare.

In Latin America, Brazil's Belo Monte Dam represents one of the largest hydropower projects in the world. Located on the Xingu River in the Amazon Basin, this massive undertaking has a capacity of over 11,000 megawatts, making it a cornerstone of Brazil's renewable energy strategy. Despite its scale, the Belo Monte project faced significant challenges, including environmental concerns and social opposition. To address these issues, developers implemented comprehensive environmental and social management plans, which included measures to protect biodiversity, support indigenous communities, and mitigate the project's impacts. Through ongoing dialogue and collaboration with stakeholders, the Belo Monte project has demonstrated the potential for large-scale hydropower to contribute to sustainable development while respecting local ecosystems and cultures.

In Southeast Asia, Laos' Nam Theun 2 Hydropower Project exemplifies the potential of hydropower to drive economic growth and poverty reduction. With a capacity of 1,070 megawatts, this project is a key component of Laos' strategy to become a regional energy exporter. The Nam Theun 2 project was developed through a public-private partnership, with support from international financial institutions and development agencies. A significant portion of the project's revenue is allocated to social and environmental programs, including education, healthcare, and biodiversity conservation. By integrating sustainable development goals into its operations, the Nam Theun 2 project has not only enhanced energy access but also contributed to the broader development of Laos.

In the Pacific Islands, Fiji's Nadarivatu Hydropower Station showcases the potential of small-scale hydropower to enhance energy resilience in remote regions. Commissioned in 2012, this

41-megawatt project was developed to reduce Fiji's dependence on imported diesel and increase its renewable energy capacity. The Nadarivatu project was designed with a focus on environmental sustainability, incorporating measures to protect local ecosystems and minimize its ecological footprint. By providing a reliable source of clean energy, the project has improved energy security and reduced greenhouse gas emissions, supporting Fiji's commitment to climate change mitigation.

In Africa, Ethiopia's Gilgel Gibe III Dam highlights the transformative impact of hydropower on national development. With a capacity of 1,870 megawatts, this project is a critical component of Ethiopia's strategy to become a regional energy hub. The Gilgel Gibe III project was developed with a focus on maximizing local benefits, including job creation, infrastructure development, and social programs. By prioritizing local workforce development and community engagement, the project has not only enhanced energy access but also contributed to economic growth and poverty reduction in the region.

In Central Asia, Tajikistan's Rogun Dam represents a bold vision for energy independence and regional cooperation. Once completed, this project will be the tallest dam in the world, with a capacity of over 3,600 megawatts. The Rogun project is a key component of Tajikistan's strategy to harness its abundant water resources and become a major energy exporter. Despite its scale, the project has faced significant challenges, including geopolitical tensions and environmental concerns. To address these issues, developers have engaged in ongoing dialogue with neighboring countries and implemented comprehensive environmental and social management plans. By fostering regional cooperation and prioritizing sustainable development, the Rogun project has the potential to transform Tajikistan's energy landscape and contribute to regional stability.

These case studies illustrate the diverse ways in which hydropower can be harnessed to drive development and enhance energy access in developing regions. While each project is unique, they share common themes of collaboration, innovation, and a commitment to sustainability. By learning from these success stories, other communities can develop their own hydropower projects that are tailored to their specific needs and contexts.

The future of hydropower in developing regions is promising, with significant potential for growth and innovation. As technology continues to advance and costs decline, hydropower is becoming an increasingly attractive option for communities seeking to transition to renewable energy. By leveraging supportive policies and incentives, communities can overcome financial and regulatory barriers, ensuring that the benefits of hydropower are accessible to all.

As the world continues to grapple with the challenges of climate change and energy security, hydropower offers a sustainable and equitable solution. By harnessing the power of water, communities can reduce their reliance on fossil fuels, decrease greenhouse gas emissions, and create a cleaner, healthier planet for future generations. Through continued innovation, collaboration, and commitment, hydropower can play a pivotal role in shaping a sustainable energy future.

Chapter 4: Geothermal Energy: Tapping Earth's Heat for All

Basics of Geothermal Energy and Its Accessibility Potential

Geothermal energy, a powerful yet often underutilized resource, offers immense potential for sustainable energy production. By tapping into the Earth's natural heat, geothermal systems can provide a reliable and continuous source of energy, making them an attractive option for regions seeking to diversify their energy mix and reduce reliance on fossil fuels. Understanding the basics of geothermal energy and its accessibility potential is crucial for communities and policymakers aiming to harness this resource effectively.

At its core, geothermal energy is derived from the heat stored beneath the Earth's surface. This heat originates from the planet's formation and the radioactive decay of minerals, creating a vast reservoir of thermal energy. Geothermal systems exploit this heat by extracting steam or hot water from underground reservoirs and using it to generate electricity or provide direct heating. The process is both efficient and sustainable, as the Earth's heat is continuously replenished.

One of the primary advantages of geothermal energy is its ability to provide a stable and consistent power supply. Unlike solar or wind energy, which are subject to weather variability, geothermal systems can operate continuously, offering a reliable source of baseload power. This reliability is particularly valuable for regions with high energy demands or limited access to other renewable resources. By integrating geothermal energy into their energy mix, communities can enhance energy security and reduce their carbon footprint.

The accessibility of geothermal energy depends on several factors, including geological conditions, technological capabilities, and economic considerations. Geothermal resources are typically found in regions with high tectonic activity, such as the Pacific Ring of Fire, where the Earth's crust is thin and heat can easily reach the surface. However, advances in technology have expanded the potential for geothermal development beyond these traditional hotspots, enabling the exploration of deeper and less accessible resources.

Geothermal energy systems can be broadly categorized into three types: dry steam, flash steam, and binary cycle. Dry steam plants, the simplest and oldest type, use steam directly from geothermal reservoirs to drive turbines and generate electricity. Flash steam plants, the most common type, extract high-pressure hot water from the ground and convert it into steam by reducing the pressure. Binary cycle plants, which are suitable for lower-temperature resources, use a secondary fluid with a lower boiling point than water to transfer heat and generate electricity. Each system has its own advantages and limitations, and the choice of technology depends on the specific characteristics of the geothermal resource.

The development of geothermal energy projects involves several key steps, beginning with the identification and assessment of potential sites. Geologists and engineers conduct surveys and studies to evaluate the geothermal resource, considering factors such as temperature, pressure, and permeability. Once a suitable site is identified, exploratory drilling is conducted to confirm the resource's viability. This phase is critical, as it determines the project's feasibility and potential output.

Community engagement and participation are essential components of successful geothermal projects. By involving local residents in the planning and decision-making process, developers can ensure that projects are aligned with community needs and values. This participatory approach fosters a sense of ownership and empowerment, as communities take an active role in shaping their energy future. Community meetings, workshops, and training sessions provide valuable opportunities for residents to learn about the benefits of geothermal energy and contribute their insights and ideas.

Environmental considerations are an important aspect of geothermal development. While geothermal energy is generally considered environmentally friendly, it can have localized impacts, such as land subsidence, water usage, and the release of greenhouse gases. To mitigate these impacts, developers can implement measures such as reinjecting used geothermal fluids back into the reservoir, using closed-loop systems to minimize emissions, and conducting thorough environmental assessments to identify and address potential risks.

The economic benefits of geothermal energy extend beyond energy generation. By providing a reliable source of electricity and heat, geothermal projects can support local businesses, create jobs, and stimulate economic growth. For example, access to geothermal energy can enable the establishment of industries such as agriculture, aquaculture, and tourism, which can enhance the community's economic resilience and prosperity. Additionally, the construction, operation, and maintenance of geothermal plants require skilled labor, creating employment opportunities for local residents.

Financing is a critical consideration for geothermal projects, particularly in resource-constrained communities. Various funding options are available, including government grants, loans, and private investment. Community-based financing models, such as cooperatives or crowdfunding, can also be effective in mobilizing resources and ensuring that the benefits of the project are shared equitably. By pooling resources and sharing risks, communities can overcome financial barriers and achieve energy independence.

Education and capacity-building initiatives are vital for the long-term success of geothermal projects. By providing training and resources, communities can develop the skills and knowledge needed to operate and maintain their systems. Educational programs can raise awareness about the benefits of renewable energy and inspire the next generation of energy leaders. By investing in education, communities can build a foundation for sustainable development and innovation.

The integration of geothermal energy with other renewable energy sources can enhance the resilience and reliability of local energy systems. Hybrid systems that combine geothermal with solar or wind energy can provide a more stable and consistent power supply, particularly in regions with variable weather patterns. By diversifying their energy sources, communities can reduce their dependence on fossil fuels and increase their energy security.

The future of geothermal energy is promising, with significant potential for growth and innovation. As technology continues to advance and costs decline, geothermal energy is becoming an increasingly attractive option for communities seeking to transition to renewable energy. By leveraging supportive policies and incentives, communities can overcome financial and regulatory barriers, ensuring that the benefits of geothermal energy are accessible to all.

As the world continues to grapple with the challenges of climate change and energy security, geothermal energy offers a sustainable and equitable solution. By harnessing the Earth's natural heat, communities can reduce their reliance on fossil fuels, decrease greenhouse gas emissions, and create a cleaner, healthier planet for future generations. Through continued innovation, collaboration, and commitment, geothermal energy can play a pivotal role in shaping a sustainable energy future.

Small-Scale Geothermal Solutions for Local Use

Small-scale geothermal solutions offer a promising avenue for local communities to harness renewable energy in a sustainable and efficient manner. These systems, designed to provide heating, cooling, and electricity on a smaller scale, are particularly well-suited for residential, commercial, and agricultural applications. By tapping into the Earth's natural heat, small-scale geothermal solutions can significantly reduce energy costs, decrease reliance on fossil fuels, and contribute to environmental sustainability.

The concept of small-scale geothermal energy revolves around the use of ground-source heat pumps (GSHPs) and direct-use applications. GSHPs are a versatile technology that can be used for both heating and cooling. They operate by transferring heat between the ground and a building, taking advantage of the relatively constant temperature of the Earth's subsurface. In the winter, the system extracts heat from the ground and transfers it indoors, while in the summer, it reverses the process, removing heat from the building and dissipating it into the ground. This efficient exchange of heat makes GSHPs an attractive option for reducing energy consumption and lowering utility bills.

Direct-use geothermal applications involve the direct extraction of heat from geothermal resources for various purposes, such as greenhouse heating, aquaculture, and industrial processes. These applications are particularly beneficial in regions with accessible geothermal resources, where the Earth's heat can be utilized without the need for complex infrastructure. By leveraging direct-use geothermal energy, communities can enhance agricultural productivity, support local industries, and improve overall quality of life.

The implementation of small-scale geothermal solutions begins with a thorough assessment of the local geothermal potential. This involves evaluating the geological conditions, such as soil composition, groundwater availability, and temperature gradients. By conducting detailed site assessments, developers can determine the feasibility of geothermal projects and identify the most suitable technologies for the specific location. This careful planning ensures that the systems are optimized for efficiency and performance.

Community engagement plays a crucial role in the successful deployment of small-scale geothermal solutions. By involving local residents in the planning and decision-making process, developers can ensure that projects align with community needs and priorities. This participatory approach fosters a sense of ownership and empowerment, as residents take an active role in shaping their energy future. Community meetings, workshops, and training sessions provide valuable opportunities for residents to learn about the benefits of geothermal energy and contribute their insights and ideas.

The environmental benefits of small-scale geothermal solutions are significant. By reducing reliance on fossil fuels, these systems help decrease greenhouse gas emissions and mitigate climate change. Additionally, geothermal energy is a clean and renewable resource, with minimal environmental impact compared to conventional energy sources. By adopting geothermal solutions, communities can contribute to a more sustainable and resilient energy future.

Economic considerations are an important aspect of small-scale geothermal development. While the initial investment for geothermal systems can be higher than traditional heating and cooling systems, the long-term savings on energy costs can offset these expenses. Additionally, various incentives and financing options are available to support the adoption of geothermal technologies. Government grants, tax credits, and low-interest loans can help reduce the financial burden and make geothermal solutions more accessible to a wider range of communities.

Education and capacity-building initiatives are essential for the long-term success of small-scale geothermal projects. By providing training and resources, communities can develop the skills and knowledge needed to operate and maintain their systems. Educational programs can raise awareness about the benefits of renewable energy and inspire the next generation of energy leaders. By investing in education, communities can build a foundation for sustainable development and innovation.

The integration of small-scale geothermal solutions with other renewable energy sources can enhance the resilience and reliability of local energy systems. Hybrid systems that combine geothermal with solar or wind energy can provide a more stable and consistent power supply, particularly in regions with variable weather patterns. By diversifying their energy sources, communities can reduce their dependence on fossil fuels and increase their energy security.

The future of small-scale geothermal solutions is promising, with significant potential for growth and innovation. As technology continues to advance and costs decline, geothermal energy is becoming an increasingly attractive option for communities seeking to transition to renewable energy. By leveraging supportive policies and incentives, communities can overcome financial and regulatory barriers, ensuring that the benefits of geothermal energy are accessible to all.

As the world continues to grapple with the challenges of climate change and energy security, small-scale geothermal solutions offer a sustainable and equitable solution. By harnessing the Earth's natural heat, communities can reduce their reliance on fossil fuels, decrease greenhouse gas emissions, and create a cleaner, healthier planet for future generations. Through continued innovation, collaboration, and commitment, small-scale geothermal solutions can play a pivotal role in shaping a sustainable energy future.

Overcoming Technical and Financial Barriers to Geothermal Adoption

Geothermal energy, with its promise of sustainable and reliable power, faces several hurdles that must be addressed to unlock its full potential. Overcoming technical and financial barriers is crucial for the widespread adoption of geothermal systems, particularly in regions where this resource remains untapped. By understanding these challenges and implementing strategic solutions, communities and developers can pave the way for a geothermal future.

Technical barriers to geothermal adoption often stem from the complexities of resource identification and extraction. Unlike solar or wind energy, which are readily available and visible, geothermal resources are hidden beneath the Earth's surface. This necessitates detailed geological surveys and exploratory drilling to assess the viability of a site. The process can be time-consuming and costly, requiring specialized expertise and equipment. To mitigate these challenges, advancements in geophysical imaging and remote sensing technologies are being leveraged to improve the accuracy and efficiency of resource assessments. By investing in research and development, the geothermal industry can enhance its ability to identify promising sites and reduce the risks associated with exploration.

Another technical challenge lies in the variability of geothermal resources. The temperature, pressure, and chemical composition of geothermal fluids can vary significantly from one location to another, affecting the design and operation of geothermal systems. To address this variability, developers must tailor their technologies to the specific characteristics of each resource. This may involve selecting the appropriate type of geothermal plant, such as dry steam, flash steam, or binary cycle, and optimizing the system's components for maximum efficiency. By adopting a flexible and adaptive approach, developers can ensure that geothermal systems are well-suited to their unique environments.

The integration of geothermal energy into existing energy systems presents additional technical challenges. Geothermal plants often require connection to the electrical grid, which may necessitate upgrades to transmission infrastructure. In remote or rural areas, where grid access is limited, off-grid or microgrid solutions may be more appropriate. These systems can provide localized energy generation and distribution, enhancing energy resilience and reducing transmission losses. By exploring innovative grid integration strategies, communities can maximize the benefits of geothermal energy and support the transition to a more sustainable energy system.

Financial barriers are a significant obstacle to geothermal adoption, particularly in regions with limited access to capital. The high upfront costs of geothermal projects, including exploration, drilling, and plant construction, can be prohibitive for many communities and developers. To overcome these financial challenges, a range of funding mechanisms and incentives can be employed. Government grants, tax credits, and low-interest loans can help reduce the financial burden and make geothermal projects more attractive to investors. Additionally, public-private partnerships and community-based financing models, such as cooperatives or crowdfunding, can mobilize resources and share risks, ensuring that the benefits of geothermal energy are accessible to a wider range of stakeholders.

The economic viability of geothermal projects is closely linked to their scale and efficiency. Small-scale geothermal systems, such as ground-source heat pumps, can offer a cost-effective solution for residential and commercial applications. By providing heating and cooling at a lower cost than traditional systems, these technologies can deliver significant savings over their lifetime. For larger-scale projects, economies of scale can be achieved by developing multiple geothermal plants in a region, sharing infrastructure and resources. By optimizing the scale and efficiency of geothermal systems, developers can enhance their economic competitiveness and attract investment.

Policy and regulatory frameworks play a crucial role in facilitating geothermal adoption. Supportive policies, such as renewable energy targets, feed-in tariffs, and streamlined permitting processes, can create a favorable environment for geothermal development. By providing clear and consistent guidelines, governments can reduce the uncertainty and complexity associated with project development, encouraging investment and innovation. Additionally, international cooperation and knowledge-sharing can help disseminate best practices and lessons learned, accelerating the adoption of geothermal energy worldwide.

Education and capacity-building initiatives are essential for overcoming technical and financial barriers to geothermal adoption. By providing training and resources, communities can develop the skills and knowledge needed to operate and maintain geothermal systems. Educational programs can raise awareness about the benefits of renewable energy and inspire the next generation of energy leaders. By investing in education, communities can build a foundation for sustainable development and innovation.

The integration of geothermal energy with other renewable energy sources can enhance the resilience and reliability of local energy systems. Hybrid systems that combine geothermal with solar or wind energy can provide a more stable and consistent power supply, particularly in regions with variable weather patterns. By diversifying their energy sources, communities can reduce their dependence on fossil fuels and increase their energy security.

The future of geothermal energy is promising, with significant potential for growth and innovation. As technology continues to advance and costs decline, geothermal energy is becoming an increasingly attractive option for communities seeking to transition to renewable energy. By leveraging supportive policies and incentives, communities can overcome financial and regulatory barriers, ensuring that the benefits of geothermal energy are accessible to all.

As the world continues to grapple with the challenges of climate change and energy security, geothermal energy offers a sustainable and equitable solution. By harnessing the Earth's natural heat, communities can reduce their reliance on fossil fuels, decrease greenhouse gas emissions, and create a cleaner, healthier planet for future generations. Through continued innovation, collaboration, and commitment, geothermal energy can play a pivotal role in shaping a sustainable energy future.

The Role of Geothermal Energy in Energy Security

Geothermal energy, a cornerstone of renewable energy strategies, plays a pivotal role in enhancing energy security. As nations strive to reduce their dependence on fossil fuels and transition to sustainable energy systems, geothermal energy offers a reliable and continuous power source that can significantly bolster energy resilience. By understanding the unique attributes of geothermal energy and its contributions to energy security, communities and policymakers can make informed decisions to integrate this resource into their energy portfolios.

At the heart of geothermal energy's appeal is its ability to provide a stable and consistent power supply. Unlike solar and wind energy, which are subject to fluctuations due to weather conditions, geothermal energy is available 24/7, regardless of external factors. This reliability makes it an ideal candidate for baseload power generation, ensuring a steady supply of electricity to meet the demands of homes, businesses, and industries. By incorporating geothermal energy into their energy mix, countries can reduce their vulnerability to energy shortages and price volatility, enhancing overall energy security.

Geothermal energy's contribution to energy security extends beyond its reliability. It also offers a domestic energy source that can reduce reliance on imported fuels. For many countries, energy imports represent a significant portion of their energy consumption, exposing them to geopolitical risks and market fluctuations. By tapping into domestic geothermal resources, nations can decrease their dependence on foreign energy supplies, strengthening their energy independence and resilience. This shift not only enhances national security but also supports economic stability by reducing the impact of global energy price swings.

The environmental benefits of geothermal energy further underscore its role in energy security. As a clean and renewable resource, geothermal energy produces minimal greenhouse gas emissions compared to fossil fuels. By reducing carbon emissions, geothermal energy contributes to climate change mitigation efforts, aligning with global sustainability goals. This environmental advantage is increasingly important as countries face growing pressure to transition to low-carbon energy systems. By investing in geothermal energy, nations can meet their climate commitments while ensuring a secure and sustainable energy future.

Geothermal energy's potential to support energy security is particularly pronounced in regions with abundant geothermal resources. Countries located along tectonic plate boundaries, such as those in the Pacific Ring of Fire, have significant geothermal potential that can be harnessed for power generation. By developing these resources, nations can create a diversified energy portfolio that enhances resilience and reduces reliance on a single energy source. This diversification is crucial for mitigating the risks associated with energy supply disruptions and ensuring a stable energy supply.

The integration of geothermal energy into existing energy systems requires careful planning and coordination. Infrastructure development, such as the construction of geothermal plants and transmission lines, is essential for harnessing geothermal resources effectively. Additionally, regulatory frameworks and policies must be in place to support geothermal development and facilitate investment. By creating a conducive environment for geothermal energy, governments can attract private sector participation and accelerate the deployment of geothermal projects.

Community engagement is a vital component of successful geothermal energy development. By involving local communities in the planning and decision-making process, developers can ensure that projects align with community needs and priorities. This participatory approach fosters a sense of ownership and empowerment, as residents take an active role in shaping their energy future. Community support is essential for overcoming potential challenges and ensuring the long-term success of geothermal projects.

Education and capacity-building initiatives are crucial for maximizing the benefits of geothermal energy. By providing training and resources, communities can develop the skills and knowledge needed to operate and maintain geothermal systems. Educational programs can raise awareness about the benefits of renewable energy and inspire the next generation of energy leaders. By investing in education, communities can build a foundation for sustainable development and innovation.

The economic benefits of geothermal energy extend beyond energy security. By providing a reliable source of electricity and heat, geothermal projects can support local businesses, create jobs, and stimulate economic growth. For example, access to geothermal energy can enable the establishment of industries such as agriculture, aquaculture, and tourism, which can enhance the community's economic resilience and prosperity. Additionally, the construction, operation, and maintenance of geothermal plants require skilled labor, creating employment opportunities for local residents.

The integration of geothermal energy with other renewable energy sources can enhance the resilience and reliability of local energy systems. Hybrid systems that combine geothermal with solar or wind energy can provide a more stable and consistent power supply, particularly in regions with variable weather patterns. By diversifying their energy sources, communities can reduce their dependence on fossil fuels and increase their energy security.

The future of geothermal energy is promising, with significant potential for growth and innovation. As technology continues to advance and costs decline, geothermal energy is becoming an increasingly attractive option for communities seeking to transition to renewable energy. By leveraging supportive policies and incentives, communities can overcome financial and regulatory barriers, ensuring that the benefits of geothermal energy are accessible to all.

As the world continues to grapple with the challenges of climate change and energy security, geothermal energy offers a sustainable and equitable solution. By harnessing the Earth's natural heat, communities can reduce their reliance on fossil fuels, decrease greenhouse gas emissions, and create a cleaner, healthier planet for future generations. Through continued innovation, collaboration, and commitment, geothermal energy can play a pivotal role in shaping a sustainable energy future.

Future Directions for Inclusive Geothermal Development

Geothermal energy stands at the forefront of the renewable energy revolution, offering a sustainable and reliable power source that can be harnessed across diverse geographical landscapes. As the world increasingly turns to renewable energy to combat climate change and ensure energy security, the future of geothermal development must be inclusive, equitable, and innovative. By focusing on inclusive geothermal development, we can ensure that the benefits of this resource are accessible to all communities, regardless of their size, location, or economic status.

One of the key aspects of inclusive geothermal development is expanding access to geothermal resources in underserved and remote areas. Many regions with significant geothermal potential remain untapped due to geographical, technical, or financial barriers. To address these challenges, targeted efforts are needed to identify and develop geothermal resources in these areas. This may involve conducting detailed geological surveys, investing in advanced exploration technologies, and providing financial incentives to attract investment. By prioritizing the development of geothermal resources in underserved regions, we can ensure that all communities have the opportunity to benefit from this clean and sustainable energy source.

Community engagement and participation are essential components of inclusive geothermal development. By involving local residents in the planning and decision-making process, developers can ensure that projects align with community needs and priorities. This participatory approach fosters a sense of ownership and empowerment, as residents take an active role in shaping their energy future. Community meetings, workshops, and training sessions provide valuable opportunities for residents to learn about the benefits of geothermal energy and contribute their insights and ideas. By building strong partnerships with local communities, developers can create projects that are socially and environmentally responsible.

Education and capacity-building initiatives are crucial for maximizing the benefits of geothermal energy. By providing training and resources, communities can develop the skills and knowledge needed to operate and maintain geothermal systems. Educational programs can raise awareness about the benefits of renewable energy and inspire the next generation of energy leaders. By investing in education, communities can build a foundation for sustainable development and innovation. This focus on education ensures that the benefits of geothermal energy are not only realized in the present but also sustained for future generations.

The integration of geothermal energy with other renewable energy sources can enhance the resilience and reliability of local energy systems. Hybrid systems that combine geothermal with solar or wind energy can provide a more stable and consistent power supply, particularly in regions with variable weather patterns. By diversifying their energy sources, communities can reduce their dependence on fossil fuels and increase their energy security. This integrated approach to renewable energy development ensures that communities can adapt to changing energy needs and environmental conditions.

Innovative financing models are essential for supporting inclusive geothermal development. The high upfront costs of geothermal projects can be a significant barrier for many communities, particularly those with limited access to capital. To overcome these financial challenges, a range of funding mechanisms and incentives can be employed. Government grants, tax credits, and low-interest loans can help reduce the financial burden and make geothermal projects more attractive to investors. Additionally, public-private partnerships and community-based financing models, such as cooperatives or crowdfunding, can mobilize resources and share risks, ensuring that the benefits of geothermal energy are accessible to a wider range of stakeholders.

Policy and regulatory frameworks play a crucial role in facilitating inclusive geothermal development. Supportive policies, such as renewable energy targets, feed-in tariffs, and streamlined permitting processes, can create a favorable environment for geothermal development. By providing clear and consistent guidelines, governments can reduce the uncertainty and complexity associated with project development, encouraging investment and innovation. Additionally, international cooperation and knowledge-sharing can help disseminate best practices and lessons learned, accelerating the adoption of geothermal energy worldwide.

The environmental benefits of geothermal energy further underscore its role in inclusive development. As a clean and renewable resource, geothermal energy produces minimal greenhouse gas emissions compared to fossil fuels. By reducing carbon emissions, geothermal energy contributes to climate change mitigation efforts, aligning with global sustainability goals. This environmental advantage is increasingly important as countries face growing pressure to transition to low-carbon energy systems. By investing in geothermal energy, nations can meet their climate commitments while ensuring a secure and sustainable energy future.

The economic benefits of geothermal energy extend beyond energy generation. By providing a reliable source of electricity and heat, geothermal projects can support local businesses, create jobs, and stimulate economic growth. For example, access to geothermal energy can enable the establishment of industries such as agriculture, aquaculture, and tourism, which can enhance the community's economic resilience and prosperity. Additionally, the construction, operation, and maintenance of geothermal plants require skilled labor, creating employment opportunities for local residents.

The future of geothermal energy is promising, with significant potential for growth and innovation. As technology continues to advance and costs decline, geothermal energy is becoming an increasingly attractive option for communities seeking to transition to renewable energy. By leveraging supportive policies and incentives, communities can overcome financial and regulatory barriers, ensuring that the benefits of geothermal energy are accessible to all.

As the world continues to grapple with the challenges of climate change and energy security, geothermal energy offers a sustainable and equitable solution. By harnessing the Earth's natural heat, communities can reduce their reliance on fossil fuels, decrease greenhouse gas emissions, and create a cleaner, healthier planet for future generations. Through continued innovation, collaboration, and commitment, geothermal energy can play a pivotal role in shaping a sustainable energy future.

Chapter 5: Bioenergy: Turning Waste into Wealth for Communities

Overview of Bioenergy Technologies and Their Accessibility

Bioenergy technologies represent a diverse and dynamic segment of the renewable energy landscape, offering a range of solutions for converting organic materials into usable energy. These technologies harness the energy stored in biological matter, such as plants, agricultural residues, and organic waste, to produce heat, electricity, and biofuels. As the world seeks to transition to sustainable energy systems, bioenergy technologies play a crucial role in reducing reliance on fossil fuels and mitigating climate change. Understanding the various bioenergy technologies and their accessibility is essential for communities and policymakers aiming to integrate these solutions into their energy strategies.

At the core of bioenergy technologies is the concept of biomass, which refers to organic materials that can be used as fuel. Biomass can be derived from a variety of sources, including agricultural crops, forestry residues, animal manure, and municipal solid waste. The versatility of biomass makes it an attractive option for energy production, as it can be sourced locally and sustainably. By utilizing biomass, communities can reduce waste, support local economies, and decrease greenhouse gas emissions.

One of the most widely used bioenergy technologies is combustion, which involves burning biomass to produce heat and electricity. Biomass combustion systems can range from small-scale stoves and boilers for residential heating to large-scale power plants for electricity generation. These systems are particularly well-suited for regions with abundant biomass resources, such as agricultural or forestry areas. By converting biomass into energy, combustion technologies provide a reliable and renewable power source that can complement other forms of renewable energy.

Another important bioenergy technology is anaerobic digestion, a process that breaks down organic matter in the absence of oxygen to produce biogas. Biogas, primarily composed of methane and carbon dioxide, can be used as a fuel for heating, electricity generation, or transportation. Anaerobic digestion is particularly effective for managing organic waste, such as animal manure, food waste, and sewage sludge. By converting waste into energy, this technology not only reduces landfill use and methane emissions but also provides a valuable energy resource for communities.

Biofuels, derived from biomass, represent another significant category of bioenergy technologies. These liquid fuels, such as ethanol and biodiesel, can be used as alternatives to conventional gasoline and diesel in transportation. Ethanol is typically produced from sugarcane, corn, or other starch-rich crops through fermentation, while biodiesel is made from vegetable oils or animal fats through a chemical process known as transesterification. Biofuels offer a renewable and lower-carbon alternative to fossil fuels, contributing to energy security and reducing transportation-related emissions.

The accessibility of bioenergy technologies varies depending on several factors, including resource availability, technological maturity, and economic considerations. In regions with abundant biomass resources, such as agricultural or forestry areas, bioenergy technologies can be more readily adopted. However, the successful implementation of bioenergy projects requires careful planning and consideration of local conditions. This includes assessing the availability and sustainability of biomass feedstocks, evaluating the technical and economic feasibility of different technologies, and ensuring that projects align with community needs and priorities.

Economic factors play a significant role in the accessibility of bioenergy technologies. The initial investment costs for bioenergy systems can be substantial, particularly for large-scale projects. However, various funding mechanisms and incentives can help offset these costs and make bioenergy technologies more accessible. Government grants, tax credits, and low-interest loans can provide financial support for bioenergy projects, while public-private partnerships and community-based financing models can mobilize resources and share risks. By leveraging these financial tools, communities can overcome economic barriers and realize the benefits of bioenergy.

Policy and regulatory frameworks are critical for facilitating the adoption of bioenergy technologies. Supportive policies, such as renewable energy targets, feed-in tariffs, and streamlined permitting processes, can create a favorable environment for bioenergy development. By providing clear and consistent guidelines, governments can reduce the uncertainty and complexity associated with project development, encouraging investment and innovation. Additionally, international cooperation and knowledge-sharing can help disseminate best practices and lessons learned, accelerating the adoption of bioenergy technologies worldwide.

The environmental benefits of bioenergy technologies further underscore their importance in the transition to sustainable energy systems. By reducing reliance on fossil fuels, bioenergy technologies contribute to climate change mitigation efforts and align with global sustainability goals. Additionally, by utilizing waste materials and residues, bioenergy technologies can reduce landfill use and methane emissions, contributing to a cleaner and healthier environment.

Education and capacity-building initiatives are essential for maximizing the benefits of bioenergy technologies. By providing training and resources, communities can develop the skills and knowledge needed to operate and maintain bioenergy systems. Educational programs can raise awareness about the benefits of renewable energy and inspire the next generation of energy leaders. By investing in education, communities can build a foundation for sustainable development and innovation.

The integration of bioenergy technologies with other renewable energy sources can enhance the resilience and reliability of local energy systems. Hybrid systems that combine bioenergy with solar, wind, or geothermal energy can provide a more stable and consistent power supply, particularly in regions with variable weather patterns. By diversifying their energy sources, communities can reduce their dependence on fossil fuels and increase their energy security.

The future of bioenergy technologies is promising, with significant potential for growth and innovation. As technology continues to advance and costs decline, bioenergy is becoming an increasingly attractive option for communities seeking to transition to renewable energy. By leveraging supportive policies and incentives, communities can overcome financial and regulatory barriers, ensuring that the benefits of bioenergy technologies are accessible to all.

As the world continues to grapple with the challenges of climate change and energy security, bioenergy technologies offer a sustainable and equitable solution. By harnessing the energy stored in organic materials, communities can reduce their reliance on fossil fuels, decrease greenhouse gas emissions, and create a cleaner, healthier planet for future generations. Through continued innovation, collaboration, and commitment, bioenergy technologies can play a pivotal role in shaping a sustainable energy future.

Community-Based Bioenergy Projects: A Path to Energy Independence

Community-based bioenergy projects offer a transformative approach to achieving energy independence, empowering local populations to harness renewable resources while fostering economic and environmental sustainability. These projects, rooted in the principles of local engagement and resource utilization, provide communities with the tools to generate their own energy, reduce reliance on external sources, and create a resilient energy future.

The essence of community-based bioenergy projects lies in their ability to utilize locally available biomass resources. Biomass, which includes agricultural residues, forestry by-products, and organic waste, serves as a versatile feedstock for energy production. By tapping into these resources, communities can convert waste into valuable energy, reducing landfill use and minimizing environmental impact. This localized approach not only addresses waste management challenges but also turns potential liabilities into assets, creating a circular economy that benefits the entire community.

One of the most compelling aspects of community-based bioenergy projects is their potential to foster local economic development. By investing in bioenergy infrastructure, communities can create jobs and stimulate economic activity. The construction, operation, and maintenance of bioenergy facilities require skilled labor, providing employment opportunities for local residents. Additionally, by generating their own energy, communities can reduce energy costs, freeing up resources for other essential services and initiatives. This economic empowerment strengthens community resilience and enhances quality of life.

The success of community-based bioenergy projects hinges on active community involvement and collaboration. Engaging local stakeholders in the planning and decision-making process ensures that projects align with community needs and priorities. This participatory approach fosters a sense of ownership and accountability, as residents take an active role in shaping their energy future. Community meetings, workshops, and educational programs provide valuable platforms for dialogue and knowledge exchange, building trust and support for bioenergy initiatives.

Education and capacity-building are critical components of community-based bioenergy projects. By providing training and resources, communities can develop the skills and knowledge needed to operate and maintain bioenergy systems. Educational programs can raise awareness about the benefits of renewable energy and inspire the next generation of energy leaders. By investing in education, communities can build a foundation for sustainable development and innovation, ensuring that the benefits of bioenergy are realized for years to come.

The environmental benefits of community-based bioenergy projects further underscore their importance. By reducing reliance on fossil fuels, these projects contribute to climate change mitigation efforts and align with global sustainability goals. Additionally, by utilizing waste materials and residues, bioenergy projects can reduce greenhouse gas emissions and improve air quality, contributing to a cleaner and healthier environment. This environmental stewardship enhances community well-being and supports the transition to a low-carbon future.

Innovative financing models play a crucial role in supporting community-based bioenergy projects. The initial investment costs for bioenergy systems can be substantial, particularly for small communities with limited access to capital. To overcome these financial challenges, a range of funding mechanisms and incentives can be employed. Government grants, tax credits, and low-interest loans can provide financial support for bioenergy projects, while public-private partnerships and community-based financing models, such as cooperatives or crowdfunding, can mobilize resources and share risks. By leveraging these financial tools, communities can overcome economic barriers and realize the benefits of bioenergy.

Policy and regulatory frameworks are essential for facilitating the adoption of community-based bioenergy projects. Supportive policies, such as renewable energy targets, feed-in tariffs, and streamlined permitting processes, can create a favorable environment for bioenergy development. By providing clear and consistent guidelines, governments can reduce the uncertainty and complexity associated with project development, encouraging investment and innovation. Additionally, international cooperation and knowledge-sharing can help disseminate best practices and lessons learned, accelerating the adoption of bioenergy technologies worldwide.

The integration of bioenergy technologies with other renewable energy sources can enhance the resilience and reliability of local energy systems. Hybrid systems that combine bioenergy with solar, wind, or geothermal energy can provide a more stable and consistent power supply, particularly in regions with variable weather patterns. By diversifying their energy sources, communities can reduce their dependence on fossil fuels and increase their energy security. This integrated approach to renewable energy development ensures that communities can adapt to changing energy needs and environmental conditions.

The future of community-based bioenergy projects is promising, with significant potential for growth and innovation. As technology continues to advance and costs decline, bioenergy is becoming an increasingly attractive option for communities seeking to transition to renewable energy. By leveraging supportive policies and incentives, communities can overcome financial and regulatory barriers, ensuring that the benefits of bioenergy technologies are accessible to all.

As the world continues to grapple with the challenges of climate change and energy security, community-based bioenergy projects offer a sustainable and equitable solution. By harnessing the energy stored in organic materials, communities can reduce their reliance on fossil fuels, decrease greenhouse gas emissions, and create a cleaner, healthier planet for future generations. Through continued innovation, collaboration, and commitment, community-based bioenergy projects can play a pivotal role in shaping a sustainable energy future.

Economic and Environmental Benefits of Bioenergy

Bioenergy, derived from organic materials, offers a compelling solution to the dual challenges of economic development and environmental sustainability. As the world seeks to transition to renewable energy sources, bioenergy stands out for its ability to provide both economic and environmental benefits. By understanding these advantages, communities and policymakers can make informed decisions to integrate bioenergy into their energy strategies, fostering a sustainable future.

The economic benefits of bioenergy are multifaceted, impacting various sectors and stakeholders. One of the most significant advantages is job creation. The bioenergy industry encompasses a wide range of activities, from feedstock production and collection to processing and distribution. Each stage of the bioenergy supply chain requires skilled labor, creating employment opportunities in rural and urban areas alike. For instance, farmers can benefit from additional income by supplying biomass feedstocks, while technicians and engineers are needed to design, build, and maintain bioenergy facilities. This job creation not only supports local economies but also contributes to broader economic growth and stability.

Bioenergy also offers the potential for energy cost savings. By utilizing locally available biomass resources, communities can reduce their reliance on imported fossil fuels, which are often subject to price volatility and geopolitical risks. This energy independence can lead to more stable and predictable energy costs, benefiting both consumers and businesses. Additionally, bioenergy projects can provide a reliable source of heat and electricity, reducing the need for costly infrastructure investments in other energy sources. These cost savings can be reinvested in other community priorities, such as education, healthcare, and infrastructure development.

The environmental benefits of bioenergy are equally compelling. As a renewable energy source, bioenergy contributes to the reduction of greenhouse gas emissions, a critical factor in combating climate change. When biomass is used for energy, the carbon dioxide released during combustion is offset by the carbon dioxide absorbed by plants during their growth. This closed carbon cycle results in a lower net carbon footprint compared to fossil fuels. By replacing fossil fuels with bioenergy, communities can significantly reduce their carbon emissions, contributing to global efforts to mitigate climate change.

Bioenergy also plays a vital role in waste management and resource efficiency. Many bioenergy projects utilize waste materials, such as agricultural residues, forestry by-products, and organic waste, as feedstocks. By converting waste into energy, these projects reduce landfill use and minimize environmental pollution. This waste-to-energy approach not only addresses waste management challenges but also turns potential liabilities into valuable resources. By promoting resource efficiency, bioenergy projects contribute to a circular economy, where waste is minimized, and resources are used sustainably.

The integration of bioenergy with other renewable energy sources can enhance the resilience and reliability of local energy systems. Hybrid systems that combine bioenergy with solar, wind, or geothermal energy can provide a more stable and consistent power supply, particularly in regions with variable weather patterns. By diversifying their energy sources, communities can reduce their dependence on fossil fuels and increase their energy security. This integrated approach to renewable energy development ensures that communities can adapt to changing energy needs and environmental conditions.

Innovative financing models are essential for supporting the adoption of bioenergy technologies. The initial investment costs for bioenergy systems can be substantial, particularly for small communities with limited access to capital. To overcome these financial challenges, a range of funding mechanisms and incentives can be employed. Government grants, tax credits, and low-interest loans can provide financial support for bioenergy projects, while public-private partnerships and community-based financing models, such as cooperatives or crowdfunding, can mobilize resources and share risks. By leveraging these financial tools, communities can overcome economic barriers and realize the benefits of bioenergy.

Policy and regulatory frameworks play a crucial role in facilitating the adoption of bioenergy technologies. Supportive policies, such as renewable energy targets, feed-in tariffs, and streamlined permitting processes, can create a favorable environment for bioenergy development. By providing clear and consistent guidelines, governments can reduce the uncertainty and complexity associated with project development, encouraging investment and innovation. Additionally, international cooperation and knowledge-sharing can help disseminate best practices and lessons learned, accelerating the adoption of bioenergy technologies worldwide.

Education and capacity-building initiatives are critical for maximizing the benefits of bioenergy technologies. By providing training and resources, communities can develop the skills and knowledge needed to operate and maintain bioenergy systems. Educational programs can raise awareness about the benefits of renewable energy and inspire the next generation of energy leaders. By investing in education, communities can build a foundation for sustainable development and innovation, ensuring that the benefits of bioenergy are realized for years to come.

The future of bioenergy is promising, with significant potential for growth and innovation. As technology continues to advance and costs decline, bioenergy is becoming an increasingly attractive option for communities seeking to transition to renewable energy. By leveraging supportive policies and incentives, communities can overcome financial and regulatory barriers, ensuring that the benefits of bioenergy technologies are accessible to all.

As the world continues to grapple with the challenges of climate change and energy security, bioenergy offers a sustainable and equitable solution. By harnessing the energy stored in organic materials, communities can reduce their reliance on fossil fuels, decrease greenhouse gas emissions, and create a cleaner, healthier planet for future generations. Through continued innovation, collaboration, and commitment, bioenergy can play a pivotal role in shaping a sustainable energy future.